AF897726

FALL OUT.

TEN COMMON BELIEFS THAT KILL OUR BRAVEST AND THE TRUTHS THAT CAN SAVE THEM

SHAUNA 'DOC' SPRINGER, PH.D.

A preview...

You know that pit in your stomach...that gut sense that what you're hearing at a wellness conference isn't right?

Have you ever felt "trapped" in a mandatory mental health training that completely misses the mark?

Has this silent thought ever run through your mind— 'this person means well, but this would never work?'

You're not alone.

For decades, myths and false wisdom about suicide, resilience, and trauma recovery have taken us in the wrong direction.

No doubt they're well intended—but they're killing the very people they're meant to protect.

It's time to step out of line, rethink everything, and move with purpose in a new direction.

It's time to fall out.

This book exposes ten deeply rooted beliefs that have pushed wellness initiatives for our nation's protectors off course—and replaces them with hard truths, real strategies, and a range of new insights about what works.

It's not about blame—it's about breakthrough. For those who know the system is broken and refuse to accept more of the same, and for anyone who refuses to watch more heroes fall.

Because when we finally let go of false wisdom, real healing can begin.

HIDDEN IVY PRESS

FALL OUT.

TEN COMMON BELIEFS THAT KILL OUR BRAVEST
AND THE TRUTHS THAT CAN SAVE THEM

Copyright © 2026 by Shauna 'Doc' Springer, Ph.D.

All rights reserved under the Pan-American and International Copyright Convention.

This book may not be reproduced in whole or in part, except for the brief quotations embodied in critical articles or reviews, in any form or by any means, electronic or mechanical, including photocopying, recording, or by any information storage and retrieval system now known or hereinafter invented, without the permission of Shauna Springer, Ph.D.

For further information, visit www.thinlineadvisory.com

ISBN: 978-1-7368244-3-6

FALL OUT.

TEN COMMON BELIEFS THAT KILL OUR BRAVEST
AND THE TRUTHS THAT CAN SAVE THEM

HIDDEN IVY PRESS

© 2026. Shauna 'Doc' Springer, Ph.D. All rights reserved.

DEDICATION

This book is for protectors and defenders, the spouses, partners, and family members who love them, and the professionals, peers, and organizations that work to support them as they face the mental warfare that is part of their chosen path.

TABLE OF CONTENTS

FOREWORD

In the brave history of America, we have always been fortunate to have valiant men and women who answer the call to service. It has always been a generational commitment, one that has never been ignored or gambled with. One that is almost exclusive to the American culture.

With such sacrifice comes a heavy price: the burden of exposure to many unpleasant and traumatic situations that few can imagine. Often, these men and women choose to stay silent, refusing to seek help. And when the weight becomes too heavy, they resort to extreme and adverse actions to cope. Some survive. Others don't.

The reluctance to seek help is rooted in a lack of institutional trust for the providers that are supposed to help. There is also fear of retribution by employers and leaders, associated with a stigma of being seen as damaged or unreliable. While some strides have been made, we have a long way to go in getting traumatic experience care right.

I know this because the above describes my situation. The denial of need of care. The fear of damaging my reputation. The perception of weakness amongst my peers. The disregard for my family's plea to get help. The mistrust of the system. The mistake of waiting too long. We need to do better; we need radical changes to the approach of mental health.

A few years ago, I met Shauna via the publication of her book *WARRIOR*, and I immediately gravitated towards her approach to dealing with post-traumatic stress. It was not the routine procedural take on how to care for warriors and first responders. It was a visceral method to deal with a visceral issue. It was led by experience and research, not via theory. It was real.

A fact about the current state of care in America; we have an oversaturation of systems led by a high demand for services. Not enough providers. Not enough support from elected officials to explore new avenues and holistic approaches. Not enough buy-in from those who need care the most. And the go-to mitigator to deal with the above challenges: a bunch of pills that are killing our sense of existence and often leading to us taking our lives. Enough.

In this book, Shauna breaches the issue like the point-man in a close-quarters battle stack. Focused, skilled, determined, and reliable. She puts the spotlight on the real issues: myths and false wisdom about suicide, resilience, and trauma recovery. She strives to right the course to lead us in the direction of a better approach. And she does this all for a simple reason...because she cares.

I am both proud and honored to contribute to this book in the hope and purpose of creating a better system for our servants of society and freedom. They deserve better than they have now. They all need our collective help. Shauna Springer is simply leading the charge.

Doc, with you until the end, all to support the valiant few who dare go forth!

~ Ramón "CZ" Colón-López, Best Selling Author of *Carnivore Leadership: Taking Charge Instead of taking Shit!* And former Pararescueman and 4th Senior Enlisted Advisor to the Chairman of the Joint Chiefs of Staff

INTRODUCTION

Status quo

I submitted the content of this book as a presentation to a premier conference focused on preventing suicide and supporting first responder wellness. My proposal, *Ten Commonly Held Beliefs That Are Killing Police Officers*, was rejected.

It's worth pausing here. A message intended to save lives—built from years of research, firsthand experience, and clinical insight—was deemed unfit (or unimportant) for the very forum designed to protect officers. The irony is stark, but it underscores this uncomfortable truth: after decades of losing our protectors to suicide, we're still afraid to confront the beliefs that silently cost veterans and first responders their lives.

Let's be honest—there's a ton of pressure to just go along with what's always been done. If you hear the same ideas on repeat, after a while, it's easy to believe they must be true. But what if they're not? What if our "best practices" aren't working for everyone?

Maybe it's time to take a hard look at what we've accepted for years and ask if these beliefs are actually helping—or just holding us back. If we don't question the status quo, we risk missing what people really need.

Sometimes, the wrong ideas are repeatedly upheld just because nobody speaks up with a solid, credible alternative perspective.

The fallout.

This much is undeniable—*failed approaches have produced far too much fallout*.

Too much funding wasted on well-intended campaigns that do more harm than good.

Too much spent on mental health initiatives that are out of touch with protectors' needs.

Too much time and focus squandered on training based on civilian approaches.

Too much broken trust from countless acts of courage that led to career repercussions for those who protect us.

Too much focus on "resilience" and not enough focus on building the "elasticity" that leads to peak performance.

Too much focus on "wellness" (aiming to be "well" does not inspire the warrior spirit) and not enough focus on being tested, stretched and growing through, and because of, adversity.

Too much fear and "territory-guarding" that has blocked access or funding for innovative treatments that have been lifesaving for many.

Too many wellness programs that sound good on paper but exist only to "check the box."

Too many "highly qualified experts" who can't earn the trust or connect with the warrior spirit.

Too many suicide prevention leaders who have never walked someone through a tunnel of despair.

Too many people who "want to work with warriors and first responders" who do not understand that this work is not a job, but a calling, which requires its own kind of warrior spirit.

Too many of our protectors left feeling helpless, without the actionable insights they need to get traction with mental warfare.

Too many "AI solutions" that move us farther away from what dissolves shame and binds us to hope: authentic, courageous human connection.

Too many irreplaceable people who have been lost, including people some of us would have taken a bullet for.

And too many warrior wives, husbands, partners, parents, brothers and sisters who have carried an unbearable weight of grief as a result.

This is the fallout from years of failed approaches. Loss upon loss. Silence upon silence, even as every life lost screams for change.

We must rise and decisively meet the need.

Falling out.

Changing this will require us to *fall out*, to break from the grip of false beliefs. Because refusing to continually fall into the wrong thinking is how we start to get it right.

This begins with asking some difficult questions:

- *Is what we believe about supporting veterans and first responders the truth, or is it what feels comfortable?*
- *After years of heartbreak, are we ready to question the assumptions behind our wellness policies and programs?*
- *Can we accept hard truths from outsiders when insiders can't see them?*

As I asked myself these questions, I realized something important: most of the people who need this information aren't in conference halls or on committees. They're the ones suffering in silence. I know this because I get the calls that don't come to most psychologists. I know this because I've been invited into sacred circles as a trusted Doc. And the pain of loss is something I've felt in my bones alongside countless protectors.

Who this is for

So, I wrote this book for protectors and defenders, their loved ones, and those brave enough to step into the trenches of mental warfare with them.

This is for the officers, paramedics, firefighters, and dispatchers who wake up every day carrying trauma, grief, and moral injuries they cannot show.

For the veterans who are walking the wasteland of transition from the military, feeling the loss of their Tribe as an emotional amputation.

For the families who are left wondering why the strongest, bravest person they ever knew—someone they loved and depended on—was lost to suicide.

For the peers questioning everything about what it means to be resilient, and how to survive in a system that often fails them.

And for my fellow "Docs" and those who will become trusted "Docs" in the years to come.

Purpose and Structure

This book is my best effort to "translate" between the two spheres I inhabit – on one side, the organizations and professionals who support protectors and defenders, and, on the other side, the warriors and first responders who have trusted me with their private pain.

Sometimes, the translation bends towards the network of supporters, to help explain what I've observed and learned in the deep exchanges I've had with protectors.

In the final chapters of this book, the translation moves in the opposite direction. In these chapters, I speak directly to those of you with a warrior spirit, to shed light on what isn't working, and introduce new treatments, resources, and forms of support.

It's much easier for me to write a long book than a short one. But I edited this book ruthlessly, giving myself a target of 10 pages for each chapter. I want this book to fit into your hip pockets, where it can be read, re-read, and passed along. My goal is to help transform the underlying beliefs that result in devastating impacts.

Each of the next ten chapters targets a commonly held belief that is blocking progress, increasing suffering, or elevating risk of tragic outcomes. Each chapter ends with three key insights and three questions for application.

In this way, this book is designed to serve as a practical resource for those who have a direct influence on shaping the culture within their organizations and communities. Its guidance is aimed at active-duty military personnel holding positions as Senior Non-Commissioned Officers and Chief Petty Officers, as well as command staff, wellness leaders, and peer support teams within first responder agencies. Mental health clinicians, along with a wide range of individuals working in organizations that support protectors and defenders, will also find value in its pages.

Empowering Front-Line Protectors and Their Families

However, the surest path to cultural transformation depends on reaching those who serve on the front lines. Providing this resource directly to protectors and defenders lets them use these insights, reflect on their experiences, and help change beliefs and practices in their work and lives.

Furthermore, sharing these insights with the loved ones of protectors and defenders will extend the impact of this book beyond the workplace. When families gain access to this information, they're better equipped to understand the challenges faced by those they care about and can more effectively offer support. In this way, distributing the book widely not only strengthens individuals but also builds a foundation for broader cultural change through shared understanding and collective action.

Read this book and decide for yourself. Do these observations and insights bring us closer to the truths we need to act on—truths about trauma, suicide, healing, and what it really takes to support those who protect us?

Because the alternative is the same thing we've been living with for too long: silence, shame, and the loss of irreplaceable people. This is our choice: change or suffer.

1

THE BEST OF INTENTIONS

Belief #1: *"Suicide prevention awareness campaigns are a powerful way to prevent suicide."*

Each September brings a slew of slogans meant to inspire hope. Billboards declare, "It's okay not to be okay." Social media is filled with images reminding us that "You are not alone." Hashtags multiply. And while these messages are well-intentioned, they often fall painfully short, especially for those who live by the warrior's code.

For more than two decades, I've worked alongside men and women with what I call the *warrior spirit*—people who run toward danger, who shoulder the weight of protecting others, whose lives reflect a deep sense of duty and service. Every September, as the flood of "suicide prevention awareness campaigns" rolls in, I'm reminded just how far removed many of these efforts are from the world that protectors and defenders actually inhabit.

Who Are These Campaigns Really For?

If awareness campaigns are designed to reach veterans and first responders, then they're missing their mark. Because no one is more aware of suicide than they are. The veterans I work with have carried the caskets of brothers and sisters they would have died for. They've stood in dress uniforms, hands trembling, as the sound of a final salute echoes across the cemetery.

Police officers know the heartbreak too well. Within their ranks, notifications of suicide travel like shockwaves—and often it's the officer who seemed invincible, the one who kept everyone else laughing. Firefighters, paramedics, and dispatchers often encounter suicide in the rawest way possible—first on scene, first to try to save a life, and first to bear the grief when that life cannot be saved.

For these communities, suicide isn't an abstract public health issue. It's personal. It's something they've seen, touched, and carried home in silence. As one LEO put it, "I often feel like these campaigns, although admittedly very well intentioned, are more for the people watching than they are the people in the game."

The Burden of Awareness

So, when yet another awareness campaign arrives with statistics and slogans, it can feel less like support and more like exposure—like being placed under a microscope. These messages can unintentionally reinforce helplessness: *Look how bad it is. Look how many are dying.*

As a Marine once told me, "These campaigns make it sound like suicide is a runaway train—and we're the ones tied to the tracks."

The problem isn't that these campaigns care too little; it's that they understand too little. The language of public health doesn't always translate into the language of the warrior. The result is a trust gap—a cultural divide between those who serve and those who want to help.

The Weight of Guilt and Shame

Marine Corps veteran and author Steven Pressfield once wrote of "the secret shame of the warrior...the knowledge within his own heart that he could have done better, done more." That quote lives in the marrow of those who serve. It's not just poetic—it's painfully accurate.

I once sat with a firefighter who had lost his closest friend to suicide. Through tears, he whispered, "He called me the night before. I heard the phone call come in, but I was too tired to pick it up." The guilt in his voice was because of a sacred connection. Though he never could have known, he had nonetheless taken his friend's death as a personal failure.

As a dispatch director once told me, "I dread September because it is 'suicide prevention awareness month.' Every year, it just reminds me of people I've lost personally and on the job. It takes me back to how helpless I've felt on certain calls. It makes my grief so much worse."

This is the unseen battle that many awareness campaigns miss. They frame suicide prevention as a simple matter of "reaching out" or "checking in." But for warriors who measure their worth in competence, courage, and control, such messages can do real harm. When anyone, especially an outsider, tells a veteran or first responder to "just reach out," it's like telling a drowning person to swim harder—oversimplifying the problem without addressing the underlying causes.

Invisible Injuries and Systemic Wounds

The truth is that suicide among veterans and first responders doesn't arise from weakness—it grows from moral injury, trauma, betrayal, and exhaustion that go unhealed. Many carry invisible wounds: guilt for those they couldn't save, anger at systems that failed them, and disillusionment when institutions talk about "resilience" but provide little tangible support.

An EMT once described it to me this way: "They tell us to talk about our feelings, but the minute we do, we get stood down. So, we stop talking."

That's the paradox. The very culture that keeps them alive in chaos—the stoicism, composure, and toughness—can also isolate them in despair. Awareness campaigns rarely acknowledge this paradox, much less address it.

From Awareness to Action

We must shift the conversation from *awareness* to *action*. Awareness without practical insights only increases feelings of helplessness. What our protectors need is not pity or platitudes, but partnership. They need culturally informed, trust-based approaches that honor their ethos rather than challenge it. They need leaders who model vulnerability as strength, not as weakness.

Real suicide prevention starts in connection—within the tribe, within the brotherhood and sisterhood of service, and with trusted Docs who get it. Healing begins when veterans and first responders see that it's possible to struggle and still stand tall, to openly face mental warfare and still be respected, to seek counsel and still own their warrior spirit.

Considering the Impact of Awareness Campaigns

If an awareness campaign is designed to raise money to prevent suicide, it's important for organizers to consider whether the campaign's impact may unintentionally increase the very risk they're trying to reduce. While the intentions behind these campaigns are positive, there's a need for careful reflection on their actual effects.

Organizers should ask themselves if the messaging and approach of the campaign might reinforce feelings of helplessness, guilt, or responsibility among those most affected by suicide. By evaluating the potential consequences, campaigns can be designed in ways that truly support protectors and defenders—building trust and fostering connection, rather than unintentionally exacerbating risk.

The Way Forward

Awareness was never enough. It never could be.

If we truly want to reduce suicide among those who protect us, we must build bridges of trust, not towers of slogans. We must create spaces where the warrior spirit can heal without shame. And we must ensure that every campaign, every initiative, every conversation about suicide prevention speaks the language of those who have seen the worst—and still show up to serve again.

Connection is the most powerful form of prevention we have. *When We Connect, We Survive*. It's time we stop talking *about* veterans and first responders—and shift our energy and resources to how we will *stand with them*.

THREE KEY TAKEAWAYS

1) Awareness campaigns highlight the gap in cultural understanding between those with a warrior spirit and those who hope to support them.

2) Awareness campaigns unintentionally increase the burden of grief and reinforce feelings of helplessness, guilt, and responsibility for those who have died by suicide.

3) Instead of awareness campaigns, protectors and defenders need culturally informed, trust-based approaches that honor their ethos rather than challenge it.

Three questions for further reflection.

1) The last time you saw a slogan or advertisement for an awareness campaign, how did you react internally? What memories came up? Did you feel inspired or dispirited?

2) If you had the power to give your honest feedback about the impact of an awareness campaign to someone who proposes to invest funds this way, how would you explain this in your own words?

3) How would you re-design an awareness campaign that integrates the insights sharing in this chapter?

2

THE WRONG TARGET

Belief #2: *Promoting "resilience" decreases suicide risk.*

Rethinking Resilience

The word **"resilience"** gets thrown around a lot these days. Wellness programs, workshops, motivational talks—you name it. But here's a little secret I've heard over and over from warriors and first responders themselves: many of them don't actually like it.

Why? That's what we'll explore in this chapter.

Resilience Can Feel Insulting

Imagine this: you're part of a group that's already doing good, hard, meaningful work—protecting the public, showing up for your team, absorbing the impact of things most people couldn't even imagine. Now someone comes along and tells you that you need *more resilience*. How would that land for you? Kinda sucks, right?

For warriors and first responders, this can feel exactly like that. They're already brave, already strong. Telling them to "train to be resilient" can feel like an outsider looking in, doubting what they've been living and breathing every day. It doesn't honor the courage, skill, and heart they already bring to their work.

The Wrong Goal: Bounce Back or Suffer in Silence?

The concept of "being resilient" is all about how we "bounce back" from challenges and hardships. Upholding "resilience" as a goal can become interchangeable with messages like "suck it up" or "take Motrin, drink water."

In other words, for many, "resilience" has become aligned with a rigid form of stoicism - that no matter what happens, a warrior or first responder should bounce right back and may feel ashamed when he or she can't. This is based on a widely held myth - the myth of the invulnerable "hero." It's a seductive myth. It tells warriors and first responders that showing vulnerability is weakness. But the truth is simple: **we're all resilient... until we're not.**

Even an armored truck will break down if it's not maintained. Even the strongest of us are still human. And sadly, some of the strongest among us—our protectors and defenders—die by suicide. That tells us something important: resilience alone isn't enough.

Protectors and defenders don't take their lives because they're "not resilient" enough. They take their lives because they don't perceive a safe or effective path for addressing their private mental warfare. More often, they die because the system is failing them.

When "Resilience" Creates Distance

Here's another subtle trap: calling someone "resilient" can actually push them away from help. People who are on the ropes with their demons often conceal their struggles. Sometimes, those at highest risk appear to be peak performers.

This is because protectors and defenders are professionally good at compartmentalizing their hidden pain. They are operational in many cases, even while they are suffering and struggling in their private lives, wordlessly traveling down a path of despair. In many cases, those who die by suicide are not people others might expect to take their lives. This is equally true in both the first responder and military communities.

If a person with a warrior spirit is told they're resilient, they may feel ashamed to reach out when their private struggle brings them to a critical inflection point. "I should be able to handle this," they think. And that isolation is dangerous. The label meant to honor them can backfire, creating a split between how they appear and how they feel, driving them away from the support that could save their life.

Elasticity: A Better Way Forward

So, what's the alternative? Over the past several years, my colleagues and I, Dr. Mitch Javidi and Jeff Kingsfield, along with our team at MAGNUS | ONE have defined and developed the concept of **elasticity** as a better alternative to "resilience."

Elasticity is a neutral term. It conjures the image of flexibility. Elasticity requires mental, emotional, and physical adaptability, as well as building support systems that help us thrive. Elasticity can be applied to how we develop and maintain our core identity. It can be applied to leadership and peak performance. It can help us build and maintain strong relationships. And it can also be applied to how we heal from trauma.

Elasticity is What Transforms Us After Trauma

Trauma and moral injuries assault our identity, worldview, and capacity to trust, pulling us far beyond our previous psychological boundaries. Elasticity, in this context, is the capacity to be stretched and then re-form into a stable, meaningful, and connected version of ourselves—often with new strength, deeper purpose, and a more grounded sense of identity.

An elastic identity allows us to survive this painful stretching process, avoid "snapping" internally, and gradually regain coherence. We do this not by "returning to who we were," but by continually transforming into someone stronger and more whole.

The Real Source of Strength: Tribe

Elasticity depends on trusted relationships. Trauma happens in isolation. Healing happens in trusted connection.

True resilience doesn't live within a single person—it's in the **team, the crew, the Tribe**. Firefighters, police, military units—they survive, adapt, and thrive together.

When someone struggles, the support of their Tribe—whether that's family, colleagues, or close friends—is what makes recovery possible. Individual toughness is nothing compared to collective resilience. This is why, in my work, I encourage those with a warrior spirit to lean into their Tribe, not just their own grit. As my MAGNUS | ONE colleagues Tiffany Andras and Brian Ellis put it, "relying only on grit is basically suffering well."

Elasticity isn't just an internal capacity—it's supported by authentic relationships, others who "hold the line" with us, and nonjudgmental witnesses to pain. These relationships act like the elastic fibers holding us together during and after stress. They absorb shock, distribute strain, and make recovery possible. Moral injuries are often best healed in community, as the power of Tribe is greater than fear and shame.

Here's an example. The final challenge before a Marine receives the name "Marine" is called "the Crucible." A "crucible" is more than a test. It's a process of transformation under intense pressure, where what enters is not the same as what emerges.

In the Marine Corps, the Crucible represents this idea in its purest form—a relentless trial designed to strip recruits down to their core, physically exhausted, mentally strained, and emotionally tested. Hunger, fatigue, uncertainty, and adversity force individuals to confront their limits and stretch beyond what they believed possible. In that environment, adaptation is not optional; it is survival.

But Marines understand that the goal is not to endure this challenge as a solitary warrior. After weeks of continuous transformation, they've learned that their greatest power lies not in their own individual capabilities, but in how they have become seamlessly interdependent. It is this esprit de corps, this camaraderie that is (non-romantic) love by another name that fuels them. They have built deep trust through shared hardship— by the time they reach the Crucible, they have been forged in fire already.

Yet, when the Crucible ends, the final transformation is unmistakable. The moment a recruit receives the Eagle, Globe, and Anchor is not simply a ceremonial milestone—it is proof that the person who began the journey no longer exists in the same form. Through challenge and adaptation, self-interest has been burned away and replaced with purpose, commitment, and identity. The Crucible does not just teach Marines how to endure; it changes how they see themselves and their place in something larger. Like metal refined by fire, they emerge stronger, unified, and permanently marked by the process, carrying forward the values and mental resolve forged in those final miles.

Elasticity and Resilience Are Different

Elasticity goes much deeper than resilience and it's more operational in its application. To this point, one LEO said, "we discuss resilience and perseverance routinely, but it's always felt more of an attribute than a skill, and no real means of building that skill."

While resilience aligns with a trauma narrative, elasticity "shifts the narrative from merely surviving trauma to learning, adapting, and evolving as a result of it" (Javidi, 2025). Elasticity fits with Dr. Javidi's seminal work on capacity-building— not just a narrow focus on how we respond when faced with challenges.

As he explains, "Resilience may get us through a crisis, but elasticity allows us to thrive afterward, transforming challenges into opportunities for growth and improvement. Enduring constant stress without change can lead to burnout. Elasticity enables continuous adaptation, fostering sustainable well-being and high performance over the long term. Elastic individuals and organizations see change as an opportunity, not a threat. They pivot quickly, while rigid systems, focused solely on resilience, struggle in the face of disruption."

Think of it this way: resilience is about surviving, while elasticity is about thriving.

Our strongest and bravest don't need to be "more resilient." They need connection, flexibility, and a system that supports them through real challenges. Promoting elasticity over resilience gives them tools to adapt, recover, and evolve—without the shame, without the pressure to "tough it out."

Because here's the truth: ***When We Connect, We Survive—and, as my colleague, Dr. Mitch Javidi, adds, When We Adapt, We Thrive.***

THREE KEY TAKEAWAYS.

1) The push for "resilience" often translates to first responders as "be tough and just keep bouncing back from adversity no matter what comes your way."

2) Calling someone out as "resilient" can make it harder for them to get support when they struggle.

3) Elasticity is a better goal because it is a neutral term aligned with peak performance which allows individuals to build their internal capacity not just to overcome adversity, but to increase flexibility in their nervous systems, sense of identity and close relationships

   (suggested resource: *Elasticity of Identity* by Dr. Mitch Javidi, Jeff Kingsfield, and Dr. Shauna Springer, 2025).

Three questions for further reflection.

1) If you are a warrior or first responder, how do you react to someone who is not a first responder or a warrior offering to train you to become more "resilient?" If you are a supporter, how do you think this would feel for a warrior or first responder who shows up every day to do work most people can't imagine?

2) When you compare the term "elasticity" with "resilience", how do these concepts translate for you? Does one inspire positive growth more than the other? Is one more likely

to lead to social comparisons than the other (i.e., thoughts about who is more resilient than others, or less so, or how "resilient" you are in comparison to other first responders?) Does one give you "more room to move"?

3) When you consider elasticity to mean range of motion, applied to your relationships, if you were to do two things this week to expand the flexibility of how you relate to your closest loved one (a spouse, child, parent or best friend for instance), what would you do to become more elastic?

3

FIRE IN THE BASEMENT

Belief #3: *Knowing the "signs" and asking "the question" in a direct way is the best way to prevent veteran and first responder suicide.*

Thoughts of suicide do not always present as flames blazing from the windows of a building. Sometimes they are like the initially invisible fire that starts in the basement or the dark curls of smoke that hide in the walls.

Sometimes firefighters refer to fire as "The Demon" and talk about fires as though they have intent to kill. This analogy is striking to me because in many years of work with warfighters, countless warriors have used the phrase "the voice of my demon" to express their darkest, most self-destructive thoughts and feelings.

In both cases, the demons that try to kill us can be stealthy. Whether they take the form of a deadly fire or the private whisper of despair, demons are good at hiding.

Suicide risk is complex, nuanced, and in some cases, invisible to some who ultimately die by suicide. Unfortunately, our existing suicide prevention strategies are inadequate and based on civilian approaches. The ways we've been taught to assess and address risk of self-harm are failing veterans and first responders.

Standard approaches to suicide prevention implement face-valid screening measures across systems of care. Existing models urge us to "ask the question" that will supposedly identify those who are at risk. The "question" is a variant of this: *"Are you having thoughts of hurting or killing yourself?"*

Yet, in practice, this approach misses many of the people who are actually at risk.

Why is this?

This approach is not based on the true nature of risk and is not attuned to the culture, ethos, or psyche of protectors and defenders. The hard truth is this: veterans and first responders who have died by their own hand don't always know how dangerous their private mental warfare has become.

Sometimes, suicide risk sits in someone's blind spot.

To address this, I've developed a new typology of risk, based on four profiles that are common among the warriors and first responders I work with. This typology is designed to collectively move us away from helplessness and failed approaches to clarity and resolve, giving us a new way forward for keeping more protectors and defenders in the fight.

PROFILE 1: Conventionally suicidal. Someone has desire and intent to die—with self-awareness. This occurs on a spectrum of risk, from passive to immediate, acute risk.

How they show up in practice: People in this category are aware of their suicidal thinking. They would likely talk about their dark and self-destructive thoughts and impulses under three conditions: 1) a safe and trusted person were to ask them, b) they are ready to give voice to their suffering, and 3) they believe that help and healing is possible for them without permanent career consequences.

In other words, self-awareness and the presence of an inquiring therapist, peer, or loved one doesn't guarantee that we'll see a present risk of suicide. Many of my clients, not just first responders and warriors, but business leaders, athletes, sexual assault survivors, and military spouses for instance, are especially adept at hiding their private pain. They will disclose their mental warfare only when I have earned their trust and they have confidence of a hopeful path forward that will not sideline them from their job.

PROFILE 2: Self-destructive or high-risk, but not suicidal. People in this category have no intent to die, either in a self-aware or latent way. However, they may nonetheless die by an action they take, for example an overdose. In many cases, a death happens either after a prolonged struggle with addiction or as the result of accidentally taking an overdose in the first few times they take a drug. In this instance, a person is not truly suicidal, but there is no way to confirm this after the death.

How they show up in practice: No endorsement of suicide risk on standard screeners or in response to "the question" posed by current assessment approaches. An individual in the example above would spike on assessments of substance use, though, if asked by a trusted source, if they were ready to acknowledge it, and felt help was possible.

Let's take substances out of the picture to illustrate this through a second example. Another variant of a type II profile is people who take reckless risks, perhaps as part of a bi-polar disorder or thrill-seeking personality. Or someone who is spoiling for a fight and tries to get a bar brawl going for the sport of it.

Their life may be cut short due to a risk they take, but they still may not be inherently suicidal, just impulsive and/or careless about their safety. Personally, I might put the main character of the movie "Free Solo" in this category. Based on how he presented, at least from what the filmmakers showed us, Alex Honnold does not seem suicidal. He does not demonstrate a wish to die, but he does seem to have much lower activation in his innate fear system than most humans.

Honnold challenges himself to climb the 3,200-foot rock wall of El Capitan in Yosemite without safety ropes. He trains with total dedication and discipline. He aims to conquer the challenge and come out alive. If Alex had fallen during his climb, this would not seem to be related to a wish to die. He has chosen to put himself in a potentially lethal situation, but suicidal ideation does not appear to be driving self-harm risk.

PROFILE 3: Slow-burn intent. A person in this category is at risk as part of a bigger campaign of self-neglect.

How they show up in practice: Individuals in this category would probably not endorse being suicidal. That is, if you were to ask them questions about suicide risk, they won't agree that they are "suicidal" – this wouldn't quite feel right to them. If you were to ask a different type of question, though, they might agree that they have mentally "tapped out of life."

For example, perhaps they're regularly abusing substances. While the substance isn't the means of an intentional suicide attempt, maybe, as a result of prolonged use, their physical and mental defenses have become so fragile that they are fundamentally unprotective of their basic physical and safety needs. However, substance abuse isn't an essential feature of this type of presentation.

Individuals in this category might also be in very poor health, with unmanaged diabetes, for instance, or other chronic health concerns that are undertreated or unaddressed. They might be continuously chugging energy drinks despite a known heart condition and the urgent advice of their doctors to cut back on caffeine. They might be a celebrated firefighter or SWAT team member who is taking reckless risks out of a slow-burn intent. I see this quite a bit in my work with first responders, since they can get benefits for their family if they die in the line of duty, and they will be hailed as dying a hero's death.

PROFILE 4: Catalyzed intent. People in this category may quickly flip into a suicidal mode, but the risk sits in their blind spot. Often in this form of suicidality, substances play a strong role in a diathesis-stress phenomenon. Specifically, the state of intoxication both reveals and "gives oxygen" to underlying intent. Sometimes, these patients will credibly report that they woke up next to a pile of pills or a loaded firearm, with no memory of what happened. The suicidal mode and the intoxication are both altered states, so the combination of substances and the suicidal mode can be lethal.

How they show up in practice: People in this category would not endorse suicide risk on an assessment. Again, standard approaches would miss them entirely. Suicide risk will often sit in their blind spot, revealed only when their defenses are breached, usually by a substance. Sometimes defenses are breached by other things though, like public shame after discovery of an affair they've been having, a broken heart after a partner's infidelity, or the death of a loved one.

In these cases, suicide risk can materialize very quickly in the context of traumatic life changes. So, individuals in this category aren't actively "suicidal" in a discernible way, either within themselves or when probed by trained peers or treating clinicians who use conventional assessment approaches. However, the potential risk is there, in their blind spot, and when it's activated, they lack the defenses to overcome mental warfare when the demon of self-destructive intent surfaces. This can happen very quickly

- a perfect "firestorm" of challenging life events can be highly combustible.

The bottom line is this: Conventional thinking on suicide screening misses many of the people who are at risk. Most people who receive a typical "suicide prevention" screening measure would never endorse items that show them to be at risk. This is true in three out of the four categories in the typology I've developed and described above.

Better questions.

When I work with clients through my advisory capacity at THIN LINE ADVISORY, I teach wellness leaders and peer teams to ask different questions. Here are a few examples in a longer question set I've developed that can help us follow the smoke more accurately and see the suicide risk that sits in the blind spot more clearly.

1) ***Is your adrenaline system always "on?" Do you feel like you've lost the ability to get back to calm and control in your body? Do you feel unable to control your temper with those you love?***

   Here's how these questions can help:

   Many of the veterans and first responders in my life describe themselves as living in a state of chronic threat response. For many people, a critical missing piece of the progression towards suicide tracks back to a continually overactivated fight-or-flight system. Research backs this up. Researchers who tracked the psychological, psychophysiological, and behavioral features characterizing a suicide attempt as it unfolded in real-time observed continuously heightened sympathetic nervous system arousal leading up to suicide attempts (Coppersmith et al., 2024).

2) ***Do you ever catch yourself taking risks you wouldn't have taken earlier in your career?***

Here's how this question can help: Risk taking, especially when someone changes from their previous behavioral pattern, can be an indicator of underlying suicidal thoughts and feelings. Not wearing protective gear while rushing into danger can be latent suicidal behavior in a culturally "acceptable" form. Death by criminal or by the demon of an out-of-control fire is the mirror of "death by cop."

3) ***What's your sleep like lately? Ever find yourself staying up late because you don't want to wake up and do it all again?***

Here's how these questions can help: Disrupted sleep and dread of daily life are common, subtle indicators of burnout or hopelessness. When someone's daily life is lived against a backdrop of dread, we should take notice. Avoidance isn't just a sign of unaddressed trauma - it can be a sign of something that cuts much deeper - moral injuries, moral exhaustion, and loss of meaning and purpose. And this question also captures a loss of elasticity, which is associated with negative physical and mental health outcomes.

THREE KEY TAKEAWAYS.

1) Suicide risk is complex, nuanced, and in some cases, invisible to some who ultimately die by suicide. Unfortunately, our existing suicide prevention strategies, which are often reductively simplistic, miss many who are at risk.

2) When someone who has not earned solid trust boldly asks a warrior or first responder this question, "Are you having thoughts of hurting or killing yourself?" (as many are trained to do), this shows a lack of attunement with the culture, ethos, and psyche of protectors and defenders.

3) Because suicide risk often sits in the blind spot of our bravest, a new approach, and a new set of questions, becomes critical.

Three questions for further reflection.

1) Have you ever been trained to look for "warning signs" and "ask the question directly?" How did this approach land with you? Do you think this approach would work equally well across different groups of people?

2) For supporters, do you agree that in many cases, warriors and first responders are professionally good at compartmentalizing their pain? (If so, how should this insight change our approach to assessing risk?) If you are a warrior or first responder, have you ever lost a peer to

suicide when you didn't see it coming? (If so, you are not alone. We've lost irreplaceable people in part because we need a better approach that is informed by first responder culture and the warrior ethos)

3) If you think about your own mental warfare, what is a question that someone you trust could ask you to start a respectful, truthful, and ultimately liberating conversation?

4

ALTERED STATE OF CONSCIOUSNESS

Belief #4: *People who die by suicide are selfish - and pointing this out is a good deterrent to suicide.*

The Myth of Selfishness — How Shame Increases Suicide Risk

It's one of the most common statements made after a suicide: "Suicide is selfish."

It's said in anger, in confusion, in pain. Sometimes it's whispered. Sometimes it's shouted. Almost always, it comes from heartbreak. But labeling suicide as selfish—or worse, as shameful—doesn't protect people. It doesn't deter suicide. It deepens the silence that keeps veterans and first responders suffering alone.

A Hard Truth — Suicide Causes Massive Collateral Damage

At the same time, it's important to recognize a hard truth – the fact that suicide causes deep and lasting damage to loved ones, colleagues, and friends. As one suicide loss survivor put it, "it is taboo to 'guilt' people who may have survived a suicide attempt or may be contemplating suicide. But I have suffered for 8 years and counting since my own brother's death by suicide. It is my reality. I work very hard each day to be intentional and continue to live my life in the best way I can each day. Some days are better than others. I did not exist this way before his death by suicide."

And while healing from suicide loss is possible, some do not heal. With the right insights and the right support, survivors of suicide loss may heal and achieve post-traumatic growth, while others are at higher risk of suicide because they carry the burden of traumatic grief. Every birthday, every family gathering, every holiday carries the weight of trauma and grief. Suicide loss survivors go through shock and trauma as well as deep grief. Their friends may withdraw from them when they need support the most. They may be judged by others and feel like they are not the same person they once were. All of these things are true too.

How do we reconcile these two things?

The way through is to understand that while suicide causes massive collateral damage, it frequently arises from an altered state of consciousness or impaired judgment.

The Distorted Nature of the Suicidal Mind

To call suicide selfish assumes a person in suicidal crisis sees the world as we do. But they don't. Suicidal thinking is not ordinary thinking—it's an altered state of consciousness. In that mental space, the mind loops endlessly on one fundamental lie: *I am the problem. My death will be a gift to the people I love.*

Imagine someone with a severe eating disorder who weighs ninety pounds but looks in the mirror and sees someone overweight. Their perception is not just wrong—it's neurologically distorted. The same kind of distortion happens in the suicidal mode. Reality itself becomes warped by pain.

Those who have stood in that dark tunnel describe feeling mentally detached from their loved ones. It's as though their demons have taken them hostage, isolating them from the very people who could help. Just as an abuser isolates a victim from sources of safety, despair isolates a person from hope.

When someone breaks that isolation—when they speak the unspeakable—the darkness begins to lose its grip. Silence is the demon's ally. Connection is its defeat. *When We connect, We Survive.*

The Warrior Ethos has a Hidden Vulnerability

For those with a warrior ethos, this internal battle takes on an especially dangerous form. The same mindset that prizes self-sacrifice and honor can twist into something lethal when turned inward. A Marine once told me that suicide, to him, seemed like a way to protect his closest loved ones from his "weakness." In his mind, death was duty.

This echoes the ancient notion of *hara-kiri*—"death before dishonor." But in tactical terms, what's really happening is this: the enemy has breached the warrior's defenses, convincing him or her that self-destruction is victory. It's psychological infiltration. *Suicide is an insurgency of the mind*. That's why our protectors—our veterans and first responders—are often at higher risk. Their very code of courage can be weaponized against them.

Why Shame Backfires

Some leaders, campaigns, and even peers believe that casting suicide as shameful or cowardly might discourage it. On the surface, this seems logical: if suicide is condemned, maybe fewer will choose it. But human psychology rarely works this way.

There's a principle called *reactance theory.* It describes what happens when a person feels their freedom to choose is threatened. Tell someone forcefully that a choice is "wrong," "cowardly," or "shameful," and their instinctive reaction may be to do exactly the opposite. Shame in those with a warrior spirit often provokes rebellion. And for someone already at war with themselves, that rebellion can turn inward—with deadly results.

When we label suicide as selfish, we drive suffering people further into isolation. We teach them that their pain must stay hidden. And hidden pain is the most dangerous kind.

What People Really Mean When They Say "Suicide Is Selfish"

Over the years, I've learned that when someone calls suicide selfish, they're rarely speaking from cruelty. They're usually speaking from pain, fear, or misunderstanding. What they really mean comes down to one of three things.

1) **"I Am Hurting."**

   Suicide leaves behind devastation that defies description. I once asked a warrior what he would do if his best friend—who'd died by suicide—walked into the room. He paused for a long time and said, "I'd punch him in the face. And then I'd hug him."

   That's grief in its rawest form—rage and love tangled together. Many survivors of suicide loss convert unbearable sadness into anger because it feels safer. Saying "suicide is selfish" is often another way of saying, *I'm in pain and I don't know how to talk about it.*

2) **"I Am Afraid."**

   After a suicide, fear spreads like shrapnel. I call it *the sniper effect*—the sense that an invisible enemy has taken one of our own. Those left behind look around and think, *Who's next?* or *If someone that strong could fall, what about me?*

   This fear can paralyze entire teams or departments. It shakes the foundation of trust and safety. Calling suicide selfish is, at times, an unconscious defense against that fear—a way to create emotional distance from something that feels uncontrollable.

3) **"I Don't Understand."**

For those who've never faced suicidal thoughts, the logic of despair seems incomprehensible. How could someone with so much to live for believe their loved ones would be better off without them? But understanding the suicidal state means recognizing it as a temporary, distorted mindset—one in which the voice of despair screams louder than the voices of love.

Suicide, then, is not about selfishness. It's about *disconnection, despair, and distorted perception.* It's about a brain and heart in crisis, convinced that death is an act of mercy.

Calling to the Warrior Spirit

If we want to reach those in danger, we must speak to their warrior spirit with psychological precision. Shame doesn't activate that spirit—it shuts it down. What does activate it is *honor, connection, and purpose.*

Effective suicide prevention calls upon the same values that make someone a protector in the first place:

- **Addressing mental warfare directly is part of the warrior's path.** It's an act of courage to face what's within rather than run from it.
- **Your tribe needs you.** You are part of a team that has your back—even when you can't see it.

- **Your values are your lifeline.** The people and principles you would die for are the same ones you can *live* for.
- **Your story may help save a life.** Every time you share your struggle, you give someone else permission to survive theirs.

Turning Pain into Purpose

Grief has its own battlefield. My friend Mike Ergo, a Fallujah Marine, has found his own way to honor those he has lost—by transforming grief into motion. Mike races triathlons wearing the names of the fallen on his jersey. Every stroke, every pedal, every step is an act of remembrance. As he moves, the grief rises and falls like waves—pain turning into fuel, memory into mission. Along the way, he tells fellow athletes about the men and women whose names he carries on his racing jersey. Their stories live on, and so does the hope that others might choose to stay.

That's what healing looks like: not forgetting but carrying forward the love that never dies.

From Judgment to Understanding

If we truly want to prevent suicide among veterans and first responders, we must dismantle the language of shame. We must replace judgment with understanding and silence with connection. We must teach our protectors that their strength is not measured by how well they hide pain—but by how bravely they face it.

Shame isolates. Compassion connects. And connection saves lives.

The next time you hear someone say, "Suicide is selfish," pause before responding. Consider what they might really be saying: *I'm hurting. I'm scared. I don't understand.* Then meet them—and those still fighting their inner battles—with the grace, patience, and courage that true prevention requires.

Because the opposite of suicide is not survival. The opposite of suicide is *connection. When We Connect, We Survive.*

THREE KEY TAKEAWAYS.

1) To call suicide selfish doesn't deter suicide— this just demonstrates a failure of true understanding and/or human empathy.

2) A mind in crisis is not rational, but rather an altered state of consciousness, and this often includes distorted thinking, loss of touch with reality, mental isolation, cognitive fixation, and looping on self-destructive thoughts.

3) If we truly want to prevent suicide, we must replace moral judgment with understanding, shame with hope, and isolation with connection.

Three questions for further reflection.

1) When someone calls suicide a "selfish act" in what ways will this impact those who are grieving the loss of a loved one who has died by suicide? How can we open up space for people to walk through a grief journey without this added burden?

2) In this chapter, I shared an example of how my friend Mike Ergo turns his pain into purpose. Grief often needs a place to go— and people of action often find it helpful to grieve in active ways. What are 3 specific ideas you can think of that might help a person of action to grieve?

3) Imagine that someone very special to you is on the ropes with the voice of despair. In your words, without shaming or judging them in anyway, how would you fight back against the fundamental lie that often drives suicide attempts— the false belief that their death is somehow a gift to those they care about?

5

THE COST OF SILENCE

Belief #5: *Those who die by suicide should not be publicly memorialized.*

When Silence Speaks Volumes

Not long ago, a police department quietly changed its internal policy on line-of-duty memorials. Under the new rule, officers who died by suicide would no longer be recognized alongside those lost in the line of duty. Their names would not be read at annual memorials. Their badges would not be draped. Their portraits would not be displayed.

The decision was framed as "sensitive" and "practical." But beneath that language was something else entirely—a failure to understand what suicide really is and what such silence costs a community.

The officers we lose to suicide are not defined by the moment of their death. They are men and women who ran toward danger, who carried out their oath to protect, who absorbed trauma most people could not even imagine. They died on a different kind of battlefield—from invisible wounds sustained in the line of duty. To strip them of memorial honors is to compound their suffering—and the suffering of those they leave behind.

A Dangerous Distinction

When a department chooses not to memorialize an officer who died by suicide, it sends a quiet but devastating message: *this death is different, this individual is shameful.*

That message doesn't stay confined to the administrative memo or policy change—it spreads through the ranks like smoke. Officers who are already struggling hear it loud and clear: *If I admit my pain, if I fall, I'll be erased.*

A former chief once told me, "We just don't want to glorify suicide." I understood his fear, but his reasoning was backward. Removing remembrance doesn't prevent suicide—it multiplies the risk. Failing to honor those who die by suicide deepens stigma, reinforces secrecy, and increases despair among surviving peers. When we erase the fallen, we teach others that their lives will be judged, not remembered.

The Human Cost of Silence

Years ago, I observed what happened within an organization that lost a veteran leader to suicide after years of suicide prevention work. He had struggled silently with trauma and exhaustion, but his peers described him as "the guy everyone leaned on." When the organization chose not to include him in the annual remembrance ceremony, his closest colleagues quietly stopped attending altogether. "*If his life doesn't count*," they said, "*then what does mine mean?*"

That's the collateral damage of silence. It fractures trust, deepens the sense of isolation already haunting those on the edge, and tells families that the person they loved most died a death unworthy of honor.

Understanding the Mind in Crisis

To make moral distinctions between those who die by suicide and those who fall in the line of duty assumes that suicide is a choice made in clarity. It isn't.

It bears repeating that the suicidal state is one of distortion, confusion, and unbearable emotional pain. In that altered state, the brain convinces the sufferer that they are the problem – the liability to their family's future wellbeing or safety, that their family and their team would be better off without them. It is not selfishness or cowardice—it comes from a mind consumed by the fundamental lie that drives suicidal behavior.

Those with a warrior spirit often carry a heavy burden of years of unaddressed trauma, moral injury, and grief. Their own pain is not often the driver of suicide attempts—their unique vulnerability is how their perception of being "broken" impacts their loved ones and peers.

When we respond to suicide with exclusion, we reinforce the false belief that they were broken and unworthy of belonging in the Tribe. That is the most dangerous message we can send.

Memorials as a Path to Prevention

We often underestimate what memorials do. They are not only for the dead—they are for the living. They remind the living that each life matters, that each story counts, that service and sacrifice are honored in all their complexity.

When a veteran or first responder dies by suicide, the memorial becomes a crucial moment of *postvention*—a chance to reduce risk among those who grieve. We know that those exposed to suicide are more likely to experience suicidal thoughts themselves. In other words, a memorial can bring together *precisely* the group of people who most need to be reached with messages of hope and clarity about how to navigate their own mental warfare.

Memorials that focus on life, purpose, and connection—not on judgment or method—can be a protective factor. They give survivors permission to speak, to mourn, and to seek help. The goal is not to glorify suicide or to present a false truth. At a memorial, a range of feelings can be freely acknowledged—that we may feel rage, guilt, confusion, fear, love, and grief – all at the same time. A memorial can also be a lifesaving call to action.

Here's an example of how I might help a leader find words that are both truthful and protective of those in their ranks after a suicide loss:

> *"We've lost an irreplaceable member of our family. Many of us did not see it coming. Many of us are not just in grief – we're also angry and frustrated because we would have dropped everything if we had known how bad it was for our brother. For all who serve in the ways we do, mental warfare is part of our path. There is no avoiding it – when we're visited by thoughts of despair, let's commit to turning to the Tribe of those we love and trust so that we do not lose any more members of our family. Today, we honor our brother's service and sacrifice in the shared understanding that how a person lived, what they stood for, and who they are to us, matters more than one tragic moment in their life."*

The goal is to glorify life and service—to remind every veteran and first responder that their worth is not diminished by their struggle.

A Lesson in Leadership

A Major General in the military took this truth on board. He once said, "I can see how doing a memorial service differently for someone who died by suicide increases shame and stigma for those who are battling suicidal thoughts. You don't mean to—but you make it look like it's their fault."

Replace "soldier" with "police officer" / "paramedic" / "firefighter" / "dispatcher," and the same principle applies. Leadership sets the emotional tone of a department. When command staff choose compassion over condemnation, they teach their officers that mental health struggles are part of the human condition, not a moral failure.

Every leader in the military or in a first responder department would do well to adopt this mindset.

The Families Left Behind

For families of those who die by suicide, institutional silence is a second wound. They've already endured the unthinkable. When their loved one's name is left off the memorial wall, they're forced to grieve in private, as if their loss is somehow less worthy.

As the former Senior Advisor of Suicide Prevention and Postvention for the Tragedy Assistance Program for Survivors, I've sat with parents, spouses, siblings, and children in this position. They want to know that the person they loved is remembered not for the way they died, but for the life they lived. They want to know their loved ones' brothers and sisters in the Tribe still claim them as one of their own.

A Call to the Culture

We are long past the time when "suck it up" could serve as a leadership strategy. If we truly want to prevent suicide, we must dismantle the shame that drives it. Departments that hide or downplay suicide deaths send a message that pain is weakness. Departments that confront loss with honesty, empathy, and ceremony send a message that pain is human—and survivable.

Every department should have a clear, compassionate postvention plan that honors life and service without moral judgment. These memorials can include:

- A focus on the officer's service and impact, not the manner of death.
- Messaging that reinforces connection, belonging, and hope for those still struggling.
- Peer support and counseling made available at every ceremony.
- Leadership statements that define mental battles as a human issue, not a disciplinary one.

Why Remembering Matters

When we honor a first responder or a veteran who died by suicide, we are not condoning suicide. We are reaffirming our values: loyalty, courage, and care for one another. We are saying to every protector or defender in their circle, *You matter. You belong. There is a way through this.*

Silence isolates. Memorials connect. And connection saves lives. *When We Connect, We Survive.*

Departments that choose not to remember their fallen are not protecting their reputations—they're endangering their people. Every officer who struggles should know that their life will always mean more than their death, and that if they reach out for help, they will be met not with shame but with solidarity.

Because in the end, the measure of a department is not how it treats its strongest, but how it honors its wounded.

And how it remembers its fallen.

THREE KEY TAKEAWAYS

1) Neglecting to memorialize the life of a veteran or first responder who has died by suicide sends a quiet but devastating message: this death is different, this individual is shameful.

2) When you think about the memorial of someone who has died by suicide, you need to consider the reality that some who grieve this loss are themselves at higher risk for suicide because of the death. Memorials are a critical space for effective postvention – that is, taking actions to support healthy grieving and working to reduce the risk of losing those impacted to suicide.

3) Failing to honor the life of someone who has died by suicide does not mean that you can't be honest about the emotional impact of the loss. It's critical to balance honoring the life lived with a message that is emotionally honest and hopeful.

Three questions for further reflection.

1) How has your department or community handled the memorials of anyone you know who has died by suicide? Did the memorials you've attended balance honor, emotional honesty, and hope?

2) What impact does failing to honor the life of a first responder or veteran who has died by suicide have on an

agency/department/military unit? What impact does this have on families who are grieving a loss?

3) Imagine that you are responsible for leading the memorial of someone who has died by suicide. Let's say that you're aware that others who will attend are struggling with dark thoughts. What message would you deliver and how would you deliver this?

6

THE ILLUSION OF CONTROL

Belief #6: Every suicide can be prevented.

The Myth of Total Control

In the aftermath of suicide, one of the first things people say—especially in helping professions—is, *"Suicide is preventable."*

It's an understandable impulse. It's a declaration of hope, of compassion, of our collective refusal to give up. But for those who have lost someone to suicide, this statement can feel like a knife in the gut. Because if *all* suicides are preventable, then every suicide must also be someone's failure.

And that's not true.

The Weight of an Impossible Standard

I've sat across from first responders, veterans, and family members who've lost someone they loved—a partner, a teammate, a brother or sister in arms. I've heard them whisper through tears, "I should have seen it. I should have known." And the hardest part of these conversations is helping them understand that even when we do everything right—when we show up, listen, love, and act—there are still forces beyond our reach.

The belief that all suicides can be prevented comes from a good place. It's born from our desire to save lives, to fix what's broken. But when we elevate it to an absolute truth, it transforms from hope into guilt. It tells survivors, *If you had done more, this wouldn't have happened.*

This message doesn't just hurt families—it hurts departments, peer teams, and communities of protectors who already hold themselves to impossible standards.

The Limits of Our Control

The truth is more nuanced—and more human.

Many suicides can be prevented. With connection, treatment, and timely intervention, countless lives are saved every year. But suicide exists at the intersection of pain, biology, circumstance, and choice. It is influenced by moral injury, brain chemistry, trauma exposure, loss, identity, and timing. No one can predict with certainty when these variables will converge into a fatal decision.

I have witnessed many cases where loved ones pulled in every possible resource to help a person who was struggling with suicidal ideation. I've known people who lost a loved one after spending weeks on end staying right next to their suffering loved one. There is nothing that some of them didn't do in an attempt to intervene.

And people who have had access to wraparound, intensive care of the highest quality may still occasionally end their own lives.

To say we can prevent *all* suicides is to deny the limits of our control over another person's internal world. We can walk beside them. We can listen. We can make help accessible and remove shame. We can do everything right—and still lose them.

That is not failure. That is reality.

The Hidden Nature of the Protector's Pain

Among first responders and warriors this truth becomes even more complex. First responders and warriors are trained to perform under pressure, to compartmentalize, to control what others cannot. Their ability to manage chaos—to stay calm in the face of death and disaster—is their superpower. But it can also be their undoing.

Their *resilience* becomes their *Achilles' heel.*

First responders are professionally skilled at suppressing their emotions in order to remain operational. They can push aside exhaustion, fear, or grief and continue to function with precision. They can comfort others even when their own hearts are breaking.

This same ability allows them to hide pain so effectively that even their closest loved ones—spouses, partners, and teammates—may have no idea how deep the struggle runs. Their faces remain calm, their humor intact, their performance unimpaired. They don't "leak" distress.

While their internal defenses are crumbling, the outward signs may still be subtle, or entirely invisible.

I once worked with a firefighter who responded to every alarm with energy and focus, cracking jokes in the truck, mentoring rookies, and taking extra shifts so others could rest. No one suspected that he went home at night and sat alone in his garage, staring at the wall, thinking about how to end his life. He was operational until the day he wasn't.

That's what makes prevention so complicated in this population. Strength often camouflages suffering.

The Guilt of the Living

After a suicide, colleagues and loved ones often replay their last conversations, searching for signs they missed. *Was that a goodbye? Was that unusual?* They torture themselves with hindsight, convinced that a different word, a different question, might have changed everything.

But how do you see what someone is trained not to show?

In law enforcement, firefighting, EMS, dispatch, and the military, stoicism is more than a personality trait—it's part of the job description. Vulnerability can feel like a liability. The culture teaches protectors to "keep it together" no matter what. So even when their world is collapsing internally, they remain outwardly composed, still saving others while privately dying inside.

In my book *WARRIOR: How to Support Those Who Protect Us*, I shared the story of the boy of Sparta to highlight this truth. In brief, Spartan culture was notoriously stoic— suffering in silence was a sacred value. A young boy on his way to daily military training found a fox alongside the road. He tucked the fox into his cloak and went to class. The fox became hungry and began to eat his abdomen, but as he was a boy of Sparta, he continued to conceal the fox. The fox was only revealed darting away after he fell silently to his death, without a word of complaint.

Those with a warrior spirit may die in this way – suffering in silence until they make a tragic decision. Recognizing "signs of risk" in this community is a partial strategy at best. And conveying the message that suicide is always preventable carries the same tone.

When a colleague or loved one dies by suicide, the survivors are left wrestling with the illusion that they had more power than they did. But we cannot control what another person hides. We cannot override another's decision to stay silent.

We can only control what we do next—how we respond, how we love, how we honor them, and how we take care of ourselves.

The Only Life We Can Save with Certainty

With credit to my friend Duane France, my co-host for our 52-episode Military Times podcast "Seeking the Military Suicide Solution," there is only one suicide any of us have one hundred percent control over preventing: our own.

That truth may sound harsh, but it's actually empowering. It reminds us where our agency truly lies. We cannot single-handedly keep others alive by force of will. But we *can* build a life that keeps us connected, supported, and self-aware. We can learn the warning signs in ourselves—the moments when the weight becomes too heavy, when the world begins to narrow, when we begin to walk down a tunnel of despair. We can reach out before the demons isolate us.

This is not selfishness. It is self-preservation—the foundation of being able to serve others. Protecting our own lives is the ultimate act of courage and responsibility.

Redefining Prevention

So how should we think about prevention?

Prevention is not about omnipotence. It's about presence. It's about staying engaged, building trust, and creating conditions where people feel safe enough to speak their pain before it becomes fatal.

It's about knowing that you can't always control outcomes, but you can always control your effort—and your empathy.

Prevention is holding the tension between truth and hope:

- **Truth** that not every life can be saved, no matter how much we love or try
- **Hope** that every moment of compassion has the power to shift a trajectory we may never see

From Control to Compassion

The belief that "all suicides can be prevented" is rooted in love. But love doesn't need certainty to matter. Sometimes the bravest act of love is releasing the illusion of control, forgiving yourself for what you couldn't prevent, and channeling your pain into connection with those still here.

For the protectors and defenders among us, that connection begins with one simple truth: you are not invincible. You are human. And your humanity is not a liability—it's your lifeline.

The real victory lies not in saving *every* life, but in saving the one within your reach—your own. Because when you choose to trust, when you choose to speak, and when you choose to live, you show others that it's possible.

And that act of survival may save more lives than you'll ever know.

THREE KEY TAKEAWAYS

1) Saying that "all suicide is preventable" is a well-intended idea that can unintentionally send the message that every suicide must also be someone's failure.

2) There is no such thing as psychological X-ray vision. Especially with those who are professionally good at compartmentalizing their pain, we must acknowledge the limits of control. We cannot override another person's decision to stay silent.

3) At the same time, suicide can often be prevented with timely connection, interventions and treatment. In many cases, when the risk becomes clear, intervention can save a life, so there is also great reason to have hope that we can make all the difference in some cases.

Three questions for further reflection.

1) How do you react internally when someone says, "suicide is preventable?" Where do you go, in your mind, when you hear this message? Does this give you hope or increase negative feelings?

2) If you have lost people you care about to suicide, and you didn't see it coming, how has this shaped your perception that "suicide is preventable?"

3) In your own words, what message would you share that balances hope and reality, that acknowledges with humility the limits of our control, while inspiring those who are struggling with mental warfare that times of crisis can be overcome?

7

NOT A MEASURE OF LOVE

Belief #7: *People who are well loved won't die by suicide.*

Love Does not Equal Control of Tragic Outcomes

There's a comforting myth that circulates whenever suicide is discussed — the idea that *if someone is truly loved, they won't take their own life.*

It's a myth we cling to because it gives us a sense of safety and order. It reassures us that love, in all its power, is an impenetrable shield against despair. But the painful truth is this: **being deeply loved does not make someone immune to suicide.**

Love is powerful, yes — but it does not have unlimited stopping power.

It's Not about a Deficit of Love

I've sat with grieving parents, spouses, and teammates who are haunted by the question: *"How could this happen? They were so loved."*

They scroll through photos of laughter, shared milestones, family dinners, and holiday gatherings, trying to reconcile the warmth of those moments with the cold shock of loss.

What they're really asking is: *Why wasn't our love enough to save them?*

The answer is agonizing, but it offers freedom from misplaced guilt.

Love cannot protect someone from the inner workings of an altered mind. When a person is in deep suicidal crisis, their ability to feel love — even when it surrounds them — can collapse. Their perception becomes distorted, their sense of worth disintegrates, and they may become convinced that their death will somehow spare those they love.

In the suicidal mode, **love itself can be twisted into a weapon**.

They may believe that their loved ones would be "better off" without them, that they are a burden, that their suffering only causes pain to those around them. In this state, the very love that could tether them to life becomes distorted into a justification for leaving it.

This is a unique form of *moral pain* — the soul-deep torment that arises when a person's identity as a protector collides with the belief that they have failed those they care about most.

The Protector's Paradox

For first responders, warriors, and those wired to serve, this paradox is especially cruel. Their love is fierce, often expressed through action — service, protection, sacrifice. But that same instinct to protect can lead them into dark mental territory when they feel they can no longer fulfill that role.

Their internal dialogue often sounds like:

> *"I'm supposed to protect others, but I've become the one who causes pain."*
>
> *"I'm broken – if it were possible, it would be best for everyone if I were to be taken to an uninhabited island."*
>
> *"I'm supposed to be strong, but my weakness is putting others at risk."*

When love becomes entangled with guilt and shame, it can drive rather than deter suicidal thoughts. They're not trying to escape love — they're trying to protect the people they love from themselves.

This is why we must be careful when we tell people that "love saves lives." Love can *help* save lives, yes — but only when the person suffering still believes they are worthy of that love.

How Suicide Hijacks Connection

The suicidal mind state is a form of *psychological captivity.* A person bent on self-destruction becomes trapped in a mental loop that isolates them from the very connections that could help them survive.

In this mode, emotional reasoning overrides logic. They may be surrounded by people who love them — family, friends, colleagues — yet feel utterly alone. It's not because the love isn't real. It's because their brain is lying to them.

Imagine a radio tuned just slightly off frequency. The love is there — the signal is strong — but they can't hear it clearly. The noise of their internal pain drowns it out.

This is why it's possible for someone who is deeply loved to die by suicide. It's not about the absence of love, but the *inability to feel it* through the fog of despair.

The Dangerous Simplicity of "If They Were Loved Enough..."

Saying that love prevents suicide may sound compassionate, but it can inflict deep wounds on those left behind. It implies that if someone dies by suicide, it must be because they weren't loved enough.

This belief compounds grief with shame. Survivors of suicide loss already grapple with unbearable questions. The suggestion that their love was insufficient adds another layer of torment.

The truth is that love can coexist with unbearable pain. You can be cherished and still feel broken. You can be surrounded by people and still feel alone. You can have a family who would do anything for you — and still believe, falsely, that they'd be better off if you were gone.

From Blame to Understanding

If love alone could cure despair, none of us would lose a child, a spouse, a partner, or a brother or sister in arms to suicide. But love does something else — something equally sacred.

It anchors us in the aftermath.

It helps us rebuild meaning after the shattering.

It gives us the courage to speak openly, to challenge stigma, and to create systems of care that go beyond platitudes.

When we move from *blame* to *understanding*, we begin to see suicide not as a failure of love, but as a failure of connection in the moment it mattered most. And connection, unlike control, is something we can continually strengthen.

What Helps Instead

So, what can we do instead of assuming love is enough?

We can learn to recognize disconnection — those subtle moments when someone begins to retreat, when communication becomes shallow or mechanical. When they have difficulty sustaining eye contact. When their personality changes in unexpected ways.

When they give up fighting for the people and values we know are important to them.

We can remove shame from the conversation entirely and replace it with humility, humanity and courageous presence.

And we can stop equating love with rescue. Our love can't *fix* another person's pain, but it can open up space for their healing.

Choosing to Stay Connected

For those who have lost someone to suicide, the idea that love wasn't enough can feel like a betrayal. But perhaps the truer statement is this: *Love is never wasted.*

The love you gave mattered — even if it didn't save them. It was felt, even if it couldn't be fully believed. And in the aftermath, it's the same love that can become your reason to keep going.

Love is not a guarantee against loss, but it is the thread that binds us back to life after loss.

In the end, love's power is not in its ability to prevent every death, but in its ability to help the living stay.

THREE KEY TAKEAWAYS

1) Many who die by suicide are deeply and sincerely loved by others. Suicide often has nothing to do with a deficit of love in the life of someone who is suffering.

2) Just as a domestic violence abuser often mentally isolates his or her partner, suicidal thoughts hijack our attachment system, severing our sense of connection with those who love us and want us to stay in the fight.

3) Understanding how the warrior ethos intersects with an altered state of consciousness is critical. For example, when someone is in a suicidal mode, love itself can be twisted into a weapon, and firearms meant as tools for the protection of loved ones can be turned into weapons of self-destruction.

Three questions for further reflection.

1) Families who have lost a loved one often struggle to understand how someone surrounded by supportive people can die by suicide. In your own words, how would you integrate the insights you've learned to explain this to a bereft family member in a way that lightens their burden?

2) Shift the picture to a first responder who is grieving the loss of a peer. Now that you've read this chapter, in your own words, how would you help him or her with this kind of *moral pain* — the soul-deep torment that arises when a

person's identity as a protector collides with the belief that they have failed someone they love.

3) After losing someone precious to suicide, what are 3 practical actions we can take to demonstrate that we continue to love them while we are also grieving their loss?

8

INVISIBLE WARNING SIGNS

Belief #8: *At-risk Veterans and first responders know they're at risk.*

The Blind Spot

There's a common assumption that people who are suicidal *know* they are in danger. That they recognize the descent. That there's a moment when the warning lights go off in their minds and they can see the threat clearly — the way a pilot sees an engine fire or a police officer spots a weapon.

But that's not how it works.

For many protectors, suicide doesn't announce itself. It hides in plain sight — tucked behind stoicism, professionalism, and a deeply ingrained instinct to stay operational no matter what.

It lives in the blind spot.

The Myth of Self-Awareness

We want to believe the myth that people "see it coming" before suicide has taken hold because it comforts us. It gives us the illusion of control — the belief that we could spot the signs, or that *they* could, if only they were paying attention.

But the reality is that self-awareness is often one of the first casualties when someone enters a self-destructive mode.

The human brain, under extreme stress, shifts into survival mode. Reasoning collapses into tunnel vision. The mind stops scanning the horizon for hope and fixates on the problem directly in front of it. In this state, people often lose perspective not just on their circumstances, but on themselves.

They may not feel suicidal — they may simply feel *done.* Exhausted. Empty. Numb.

The Protector's Blind Spot

For first responders and warriors, this self-blindness is magnified by culture. They are trained to assess risk — but only external risk. Their professional survival depends on the ability to compartmentalize, to set aside fear, and to push through distress to complete the mission.

That same skill, which keeps them calm in chaos, can be deadly when the chaos is inside.

Their resilience becomes their Achilles heel.

A firefighter who can run toward a burning building without hesitation can also run headlong into personal crisis without recognizing the danger. A police officer who can walk into an armed standoff may not recognize that they themselves are standing on emotional thin ice.

They've learned to read everyone but themselves.

I've sat with countless first responders who've told me, "I didn't see it coming." They mean it literally. They didn't recognize the gradual erosion happening inside them — the subtle signs of moral exhaustion, the creeping numbness, the narrowing of focus that turns the world gray.

When they finally hit a point of crisis, it often surprises them as much as it surprises everyone else.

When the Mirror Lies

Suicidal risk is not always visible — not to others, and not even to the person experiencing it.

I once spoke with a detective who had just buried one of his closest colleagues — a man who was sharp, disciplined, and deeply admired in the department. "He wasn't like those people who die by suicide," the detective said.

I asked him gently, "What do you mean, those people?"

He hesitated. "You know... someone falling apart. Someone who gives up."

This officer, like so many others, couldn't see himself in that mirror. The idea that *he too* could one day be at risk felt incompatible with his identity as a protector.

And yet, that's how suicide hides — by convincing capable, resilient people that they're immune.

This is *a unique Achilles heel*—the same qualities that make protectors dependable under fire — composure, self-sacrifice, control— can blind them to their own suffering.

When they start to slip, they don't think, "*I'm in danger.*" They think, "*I just need to push harder.*"

The Perfect Storm

We tend to think of suicide as something that happens only to those with long-term mental illness, but that's not the full picture. Many people who die by suicide never had a diagnosable psychiatric condition.

What they had was a convergence — a perfect storm of stressors that overwhelmed their ability to cope.

It might be a moral injury after a traumatic call, a career-ending physical injury, the breakdown of a marriage, the discovery of an affair, financial pressure, or a sense of betrayal by the institution they served. Often, it's not one catastrophic event but the cumulative weight of many.

Under enough strain, even the most resilient minds can fracture.

Neuroscience tells us that when stress hormones flood the brain, they impair the very circuits responsible for insight and decision-making. Perspective collapses. The ability to "see oneself clearly" dissolves.

In this state, suicide can happen not out of long-term illness, but out of *acute overload*. It's not a decision born of clarity — it's a crisis born of distortion.

Why the Warning Signs Can Be Invisible

For most first responders, the signs of crisis are subtle because they've been conditioned to hide them. They continue to show up for shifts, crack jokes at roll call, and perform at a high level.

Externally, they're operational. Internally, they're unraveling. The picture is similar for many veterans.

When I ask colleagues after a suicide, "Did you notice any changes?" the most common response is, "No. None at all."

That's not denial — it's the nature of the beast.

The danger is that when we believe only *obvious signs* indicate risk — crying, isolation, verbal statements — we miss the silent suffering that hides beneath competence and duty.

A More Honest Story

We need to start telling a more honest story about suicide — one that acknowledges both our influence and our limits.

Yes, suicide can often be prevented when there's awareness, connection, and early intervention. But there are times when even the most vigilant peers, families, and clinicians cannot stop it.

We can be alert, compassionate, and available — but we cannot control another person's blind spots.

It bears repeating: the only suicide any of us have full control over preventing is our own.

This is not a statement of resignation; it's a call to personal vigilance. It means each of us must be willing to see ourselves as *someone who could be at risk*, no matter how strong we are or how many others depend on us.

Getting Traction with Mental Warfare

To get different outcomes, we must create cultures that see mental warfare as part of the path of being a protector or defender, not the exception that applies only to "that one person that is struggling."

Recognizing the potential for heightened risk is important – not in a fatalistic way, but in a balanced, emotionally intelligent way. Just as someone with a family history of heart disease can make different lifestyle choices as a result, the same kind of clarity can help when it comes to suicide risk. To illustrate here is an example of what a friend of mine said recently: "As a military veteran, retired police officer, and suicide loss survivor with a family history of mental health, I have accepted that I am at risk and live my life accordingly. It does not mean my life will end in suicide, rather, I use it as a means to educate others and keep myself informed and on my toes to prevent that from occurring. We must accept the risk and get traction, keep traction, and accept that mental warfare is the path to true mental elasticity."

We must be humble enough to seek new insight into how self-destructive thoughts emerge, develop, and gain momentum, even when this requires us to take guidance from unlikely sources, or extend trust to those outside our current circles.

We must understand with clarity what has real stopping power and hold this in tension with the limits of our control.

We must demonstrate what it looks like to take the fight to all forms of mental warfare without fear of losing status, trust, or belonging.

THREE KEY TAKEAWAYS.

1) We don't always know when we're in danger. Suicide risk can live in the blind spot – for any of us – but this is especially true for protectors and defenders as they are professionally good at compartmentalizing their pain.

2) Classifying those who are struggling with dark thoughts as being somehow fundamentally "different from us" only enlarges our potential blind spot.

3) We can shrink our blind spot by understanding that in a perfect storm of stress, or under certain conditions, we can all be potentially visited by self-destructive thoughts and urges.

Three questions for further reflection.

1) Have you ever had the experience of suddenly feeling the impact of cumulative stress and trauma drop on your shoulders like a thousand pounds you weren't aware you were carrying? What happens in your body and mind when this occurs?

2) How can we get better at turning within and recognizing shifts in ourselves that are taking us in the wrong direction?

3) When you become aware of the weight of your own grief, trauma or moral injuries, what do you tend to do in response? Do you isolate? Do you connect with peers or

a loved one at home? Do you have go-to strategies and a trusted Doc that can help you navigate this?

9

MISSING THE REAL REASON

Belief #9: *Courage is the main barrier to seeking help.*

The Myth of Courage

One of the most persistent myths in the world of suicide prevention is that first responders and warriors just need to have *more courage* to ask for help.

It's an easy narrative — simple, neat, and completely wrong.

The problem isn't a *deficit of courage.*

The problem is a *deficit of safety.*

Courage Is Not the Issue

No one embodies courage more than warriors and first responders. They run toward what everyone else runs from. They enter burning buildings, purposefully draw enemy fire, and approach armed suspects. They crawl into mangled vehicles and perform CPR on people who may not survive.

They have courage in quantities that most people can't imagine.

So, when someone says, "They just need to be brave enough to reach out," it betrays a profound misunderstanding of who they are and the real dangers they face when it comes to disclosing their pain.

The truth is this: **it doesn't take more courage to ask for help — it takes more safety to make it possible.**

The Real Risk Calculation

When a first responder considers reaching out for mental health support, they're not calculating bravery. They're calculating *risk.*

They ask themselves hard, practical questions:

- *"If I tell someone I'm struggling, will I lose my badge, my gun, or my assignment?"*
- *"Will my colleagues see me as weak?"*
- *"Will I be placed on leave or sidelined while they 'evaluate' me?"*
- *"Will my name end up in a report that follows me forever?"*

For many, these are not paranoid fears — they're grounded in lived experience. I've sat with first responders who did the right thing, reached out for help, and paid a professional price.

Some were placed on administrative leave for months. Others lost their clearance, their reputation, or the trust of their team. One officer told me, "Once you say you're not okay, you're never fully trusted again."

These are not cowardly people avoiding help. These are rational human beings protecting their livelihoods, their identities, and the families who depend on them.

When the system punishes vulnerability, silence is a survival strategy. And we should not ask them to be vulnerable, unless we have first created a pathway of support that is truly safe.

The Fear of Being Kicked Out of the Tribe

First responders belong to a culture built on trust — trust that your partner has your back, that your team will pull you out of the fire, that everyone in the squad car, on the rig, or in the station is part of your tribe.

To lose that belonging is to lose more than a job — it's to lose your identity.

For many, the fear of being judged, sidelined, or cut out of the tribe is stronger than the fear of death itself. I've heard countless variations of the same sentiment: *"I'd rather die than be seen as the weak link."*

That's not bravado. That's a heartbreaking reflection of how deeply belonging runs in this community. Connection is the oxygen of the protector culture — and the threat of exile suffocates people into silence.

So, when a first responder stays quiet about their mental pain, it's not because they lack courage. It's because they can't afford to lose their place in the world that defines them.

The Illusion of Access

It's easy to say "help is available" — but availability does not equal accessibility.

Yes, there are Employee Assistance Programs, mental health therapy services, peer teams, chaplains, and hotlines. But many first responders see these resources as unsafe because of confidentiality concerns, stigma, or bureaucratic entanglements.

One firefighter put it bluntly: *"I'm not going to pour my heart out to someone who reports back to my department."*

And yet, under the right conditions, these same individuals are *not reluctant at all* to open up.

When they sit with someone who truly understands their culture — someone who can meet them without judgment, rank, or threat — the floodgates open.

They are *dying* to talk (and many are dying because we haven't given them a safe place to talk).

They're hungry for connection, for safe and confidential spaces where they can tell the truth without fear of punishment or pity.

They don't need lectures about courage. They need relationships that feel trustworthy and systems that protect their confidentiality.

What Safe Looks Like

So, what does "safe" actually mean in this context?

It looks like this:

- Access to "Docs" who understand warrior culture— the gallows humor, the emotional armor, and the deep moral code that drives their work.
- Programs designed in collaboration with protectors, not imposed solely from the outside by people who have never worn the gear, held the weapon, or worked the scene.
- Strategic consultation and advising— so that command staff and wellness leaders are building programs based on the right insights.
- Peer support teams that are well-trained, with insights derived from a deep understanding of the warrior spirit.
- Services that are bound by confidentiality and publicly backed by leadership.
- Policies that allow protectors to address mental warfare head on without fear of career repercussions.

Safety must be structural, not symbolic.

Rewriting the Story

The story we tell matters. If we keep framing help-seeking as an act of courage, we imply that those who remain silent are somehow less brave.

This is not only inaccurate — it's harmful.

It reinforces shame and keeps the focus on individual strength rather than collective responsibility.

We must shift the narrative: the problem is not that our protectors are too proud or too stubborn to seek help. The problem is that we have not built systems worthy of their trust.

They are not avoiding help — they are protecting themselves from further harm.

The Way Forward

When we make it truly safe, they will come.

When they know their jobs, their status, and their families won't pay the price for their honesty, they will open up.

When we treat confidentiality as sacred and trust as the foundation — not an afterthought — the silence will start to break.

Because the truth is, they already *want* to talk. They are waiting for us to build places strong enough, and safe enough, to hold what they need to say.

So, let's stop questioning their courage. They've already proven it on every call, every shift, every scene.

It's now time for the rest of us to prove *our* courage— the courage to create systems of care worthy of the bravest among us.

THREE KEY TAKEAWAYS.

1) It doesn't take more courage to ask for help — it takes more safety to make it possible.

2) If we keep framing support-seeking as an act of courage, we imply that those who remain silent are somehow less brave.

3) First responders and veterans are not "closed off" and "unwilling to open up." They are dying to offload some of their mental burden, and some are dying because we have not created opportunities that are truly confidential and safe. The "mirage" of help that is not safe may be worse than no help at all.

Three questions for further reflection.

1) Have you ever been urged to seek help in a system of care that you sensed was not confidential? How did this change your treatment experience, if you did show up for treatment at all?

2) What is your internal reaction to this statement: "It takes courage to ask for help." What does your reaction tell you about the hidden barriers to care for many others like yourself?

3) If you were to enter a new system of care, or connect with a team of trained peers, what kinds of questions would you need answers to before you could feel safe to open up?

10

INJURED, NOT BROKEN

Belief #10: *Post Traumatic Stress is a life sentence that can never be healed.*

Post-Traumatic Stress Is Not a Life Sentence

One of the cruelest myths in the conversation about our protectors' mental health is the idea that post-traumatic stress is a life sentence...that once you've been "broken" by trauma, you are marked forever — a damaged person, labeled, and beyond help.

This belief is not only wrong — it's dangerous. It convinces people that their suffering is permanent, that seeking help is pointless, and that healing is somehow a betrayal of reality. For those living under the weight of repeated trauma, the message is devastating: "You're broken. This is who you are now."

Nothing could be further from the truth.

PTSD Is an Injury — Not a Defining Identity

Post-traumatic stress is not just a mental health battle— it's an injury to the nervous system. It's not a sign of weakness. It's not a moral failing. It's a physiological response to repeated or extreme stress that leaves the body and brain stuck in a heightened state of vigilance— what I call a "hair trigger nervous system."

Just as repeated use of a firearm can cause the trigger finger to become "frozen" into a bent position, the nervous system can become "locked" into a chronic state of fight or flight after repeated exposures to trauma. When this happens, individuals report a set of predictable challenges:

- Disrupted sleep: "I only get a few hours of sleep a night if I'm lucky."
- Sudden spikes in anxiety: "I get flooded with anxiety, even when I'm not in direct danger."
- Irritability and anger: "I go from zero to a hundred in no time at all... I just lose it and start yelling at my partner or kids. Later, I feel so ashamed."
- Difficulties concentrating: "I can't read books anymore."
- Extreme, persistent hypervigilance: "I'm so keyed up all the time that I can't even enjoy a movie in a public theater with my family."

This is not a permanent flaw in character. This is an injury to the body's alarm system — a physiological response that can, and often does, heal with the right interventions.

An Innovative Treatment Used in Special Forces Offers Real Hope

For decades, the mainstay of treating post-traumatic stress has been talk therapy and medications. While these remain crucial, a new application of a time-tested intervention has introduced an exciting option for veterans and first responders: **Dual Sympathetic Reset Stellate Ganglion Block, or DSR SGB.**

Many people—even those in mental healthcare settings, have not heard of DSR SGB, but it is not new. In fact, SGB has been a go-to strategy for many years in certain Special Forces units – to reset operators between deployments. It is a performance enhancer in this context, as it is starting to become within some of the more innovative public safety departments.

DSR SGB is a 15–20-minute procedure that targets two clusters of nerves in the neck— essentially resetting an overactive fight-or-flight system. Think of it as a "reset button" for a nervous system stuck in a state of chronic threat. Innovators have refined the technique by adding a second injection and using ultrasound guidance for greater safety and precision.

SGB has been the subject of research for approximately 100 years. A properly placed SGB injection carries no known adverse side effects. A good analogy is this: SGB has the same side effect profile as a properly placed epidural. Thousands of new mothers receive epidurals every day in birthing units across the country. The expectation with both epidurals and SGB is this: as long as the providing physician places the injection correctly, there are no long-term side effects from either procedure. This analogy is particularly apt, as SGB and epidurals during childbirth inject the same FDA-approved anesthetic medication, typically ropivacaine or bupivacaine.

Across multiple studies, 70–83% of patients experience meaningful relief, with reductions in hyper-vigilance, anxiety, and irritability that transform daily functioning. When sequenced strategically within a plan of care, DSR SGB often accelerates and enhances positive outcomes with concurrent trauma-informed talk therapy and other mind–body interventions. As a successfully treated patient once said, "SGB is the primer to the paint of seeing a Doc who gets it."

While the treatment is not yet covered by insurance, some cities, police departments and affiliated associations have started to provide ways to fund or offset the cost of care. For example, the Chicago City Council unanimously passed a landmark resolution addressing the mental health of first responders and the growing crisis among the city's emergency personnel and employees.

Led by a retired police officer and firefighter who understands firsthand the challenges first responders face, the council has now made it possible for all city employees to receive DSR SGB at Stella Mental Health through the City of Chicago's Flexible Spending Account (FSA) program. The 100 Club of Illinois covers the cost of the procedure for Illinois-based first responders if donated funds are available.

The biggest obstacle for police officers is not the cost of the DSR SGB procedure, or fears about side effects with such a minimally invasive procedure. The biggest barrier is this: "*If I get treated, will I be able to respond to threats as quickly as before, or will this slow me down?*"

The confusion stems from the original name of the procedure: stellate ganglion block. The name is based on the use of this procedure as a "nerve block" a pain treatment used for about 100 years. Contrary to its name, *SGB does not block the response to threats*. In fact, it enhances response time and operational capabilities. *SGB is used as a performance optimizer within special forces units*.

Similarly, first responders must "keep their edge." For example, many in law enforcement fear being caught "flat footed" in response to potential danger on the street. DSR SGB won't decrease operational performance, and in many cases, it will increase reaction time to threats.

With an option like DSR SGB, no one must needlessly suffer. A hair-trigger nervous system can be treated and healed. We just need to migrate what works in Special Forces units to police and fire departments, EMS companies, and veteran focused organizations across the country.

Reframing the Narrative

The myth of PTSD as a life sentence has done immense harm. It convinces people that they are irreparably damaged, that seeking help is futile, and that their suffering is permanent. This narrative isolates first responders and keeps them from exploring innovative treatments that could restore their nervous systems and their lives.

The truth is liberating: **post-traumatic stress is treatable, and recovery is possible.** It may require new interventions, connection with culturally informed Docs, and innovative peer teams. For those who have suffered in silence, the message is simple: you are not broken beyond repair. You are injured, yes, but injuries can heal. Your nervous system can reset. And your life can expand again beyond the grip of trauma.

THREE KEY TAKEAWAYS.

1) A diagnosis of PTSD is NOT a life sentence. Getting a diagnosis of PTSD does NOT mean that a veteran or first responder is broken.

2) Advances in neuroscience have now demonstrated that Post-Traumatic Stress causes a biological injury that can be seen on a brain scan and healed with effective treatment approaches.

3) DSR SGB (a modified form of Stellate Ganglion Block), a treatment used in Special Forces to support peak performance in Operators, has unique potential for addressing post-traumatic stress injuries for veterans and first responders. Here's what makes this option a great fit for many:

    - It is minimally invasive and considered a safe routine procedure
    - It does not rely on a psycho-active medicine (eliminating all risk of "popping positive" on employer drug screens)
    - It generally requires only one or two visits to a clinic
    - Downtime is only 24 hours, after which individuals can return to unrestricted activity

- It has shown great efficacy (70-83% success rate) based on treatment of many thousands of patients, and
- Research conducted on Special Forces Operators shows that it improves reaction time and neurocognitive performance

****IMPORTANT: While DSR SGB is often very effective in the short run, if you want lasting results, it is important to pair DSR SGB with sessions with a Doc who gets it.*

Why?

Because post-traumatic stress does not just injure our fight or flight system. It also causes changes in our identity and how we think, behave, and relate to others. To achieve transformative healing, these areas must also be addressed. I've taken countless veterans and first responders through treatment and observed the outcomes they achieve in comparison to those who forgo therapy. The differences I've seen are very compelling. There is no "miracle cure" or "magic bullet" — DSR SGB accelerates outcomes, but we still must do the work.**

Three questions for further reflection.

1) Have you ever felt "broken?" What would change for you if you could integrate the insight that there are new treatment options that may substantially reverse your symptoms?

2) A large study of people diagnosed with PTSD (Post Traumatic Stress Disorder) suggested that shifting from

a "Disorder" to an "Injury" model would substantially reduce stigma and make it much easier for more people to get treatment. What's your opinion? Do you think that our terms should evolve from seeing post-traumatic stress, not as a "disorder" that is a permanent condition, but rather as an injury that can be healed?

3) Have you ever heard of Stellate Ganglion Block (SGB) or Dual Sympathetic Reset Stellate Ganglion Block (DSR SGB)? Has your current health care team ever heard of this procedure? Given that this treatment has demonstrated efficacy for addressing post-traumatic stress injuries, might this be something to learn about?

11

THE BIOLOGICAL REVOLUTION

Over the past several years, I've focused my time and energy on building three things that feel good and true, deep in my own warrior spirit.

In the next three chapters, I'll share a "behind the scenes" look at these three initiatives and how each creates access to completely different forms of support than you've had in the past.

This first initiative is central to a biological revolution happening right now across the landscape of healthcare.

Stella Mental Health is an international network of clinics throughout the United States, Australia and Israel. Stella launched in 2020 and as I write this (in early 2026), Stella is treating thousands of patients every month, offering a range of innovative biological treatments.

My involvement with Stella started like this: "Doc, I keep losing my temper with my family. I don't want to, but I can't control it. Can you get me the a—hole shot?"

At the time, I was serving in the Department of Veterans Affairs. My patient – let's call him Craig (not his real name) had served as a Doc in Special Forces.

Me: "Craig, what is the a—hole shot? I've never heard of that."

Craig: "It's something we get between our combat deployments, to restore our edge since it improves reaction time. But really, a lot of us are getting it because it helps us to not be an a—hole to our wives and kids. Our pain Doc gives us a shot in the neck to get our nervous system back under control. It works immediately for most of us."

Me: "Why have I never heard of this?"

Craig: "A lot of what we do in [Special Forces] never makes its way out to most people, but we've done this for years. It's actually based on a procedure that's been used for pain conditions for nearly 100 years. If you have a pain Doc here at the VA, he's probably already doing it for things like phantom limb pain— for example, when someone loses a leg after stepping on an IED [Improvised Explosive Device]."

Me: "OK, let me look into this. I need to understand how safe and effective this is as a next step."

Craig: "I know you're looking out for me and you need to do your own research. I respect that. But I also hope you can trust me. As crazy as this may sound, I know this works, and I hope you can help me get it done."

When I think back to that conversation, I marvel that this brief exchange was a fundamental inflection point for how I think about the work of healing.

> *We ask our patients to trust us – but do we trust them?* Trust is a two-way street.
>
> *Can we see beyond the limits of our own training and expertise to the future of what is possible?*
>
> *Are we open to new treatments that force us to re-cast our entire model for how healing happens?*

What I learned did just this – it compelled me to fundamentally change my view and model for healing.

My patient was right – the Stellate Ganglion Block injection had been around for about a century. It is considered a minimally invasive procedure that is commonly used for pain conditions like shingles, complex regional pain syndrome and phantom limb pain. And in 2008, Dr. Eugene Lipov published a case study on its use for treating a patient who had suffered from trauma symptoms.

Since then, there have been several larger studies, mostly conducted by military medical providers who have used the SGB with very positive outcomes for active-duty soldiers. For instance, in one study of 250 active-duty soldiers, 100% of those treated by SGB would recommend it to a friend (McLean, 2015).

Why have I never heard of this?

Has the field of mental health been this fractured?

Have too many of us fallen into ruts of thinking, based on our training?

Have we been this territorial, guarding our patients as though we "own" them?

Have we been too fearful to trust each other, to collaborate across disciplines, to try new approaches?

The truth is that there's a universe of knowledge beyond what I learned in my training.

So, I shared this research with my colleague and friend at the VA, a highly skilled pain physician, who agreed to do this for Craig.

I was in the room with Craig when he had the procedure, standing next to him in a lead vest, as my friend used X-ray technology to place the injection in the right place.

And I sat just to his left, while his wife Celeste (not her real name) held him in a long embrace as he sobbed tears of relief just after the procedure, in deep, gulping gasps. He turned to me and said, "Doc, it feels like someone just took a thousand-pound weight off my chest." (A phrase I've now heard repeated by countless patients who do not know each other).

Over the next few months, as I wrapped up an 8-year tour of duty at the VA, Craig and I continued to meet for therapy. The entire experience was different, for both of us.

He was calm in his own body, settled, hopeful, able to hear the insights I had for him. The traction he got with these insights was radically different than before. He was no longer fighting the riptide of his own adrenaline. His relationship with his wife blossomed. They even came in for a session together just so she could thank me for "giving her back the man she married."

I realized that for years, I had been doing therapy with patients who are in the wrong state of mind - overrun with adrenaline, not sleeping, with their attention and concentration shattered.

It may help to think about this in tactical terms. You can't advance a squad until you suppress the enemy. Moving under fire requires that first step of forcing the enemies' heads down. Or if this analogy doesn't feel right, you can't secure a room until you clear it. First you must quell the chaos and then you can create stability.

Therapy is about creating stability in a system— making long-term healing possible. Therapy can help you see through the fog of mental warfare. A good Doc can help you "see the matrix" and change the patterns that were invisible to you before. Since real people can't bend before bullets hit them, we must first stop the bullets from flying at a person.

This is what SGB does – it stops the metaphorical "bullets" of the body's trauma response from flying at you. When your body and mind are restored to calm and control, you can integrate the benefits of therapy in a totally different way than before. THIS is the piece we've been missing— that first we must treat the biological injury caused by trauma, before we can heal and rebuild the system. This is the new model of care that we need to adopt— one that pairs biological treatments with culturally aligned insights.

As I wrapped up my relationships with patients at the VA, there were a few patients I was especially concerned about. Sometimes, the people who are most at risk take on a haunted look. When I say haunted, I mean that sometimes they won't make eye contact and when they do, their eyes look somehow devoid of hope. I was worried about five specific veterans under my care. I was losing sleep about how they would fare once I transferred them to a new provider. They had that haunted look. I spoke to them about SGB and all but one agreed to be treated.

All those treated had the same incredible outcome. One patient found me 5 years later and confirmed that he was deeply suicidal and that SGB had saved his life. The only patient who would not agree to be treated died by suicide about five months after I left the VA. I wonder to this day whether SGB might have saved him. These kinds of experiences leave a permanent mark on someone like me, whose purpose is to help protectors and defenders survive adversity and take the fight to mental warfare.

My understanding of the power of innovative biological treatments evolved from a private conversation into something much bigger. Stella Mental Health is an expansive partnership with responsible innovators, among them Dr. Eugene Lipov, who published the first case study on SGB for post-traumatic stress injuries. To truly understand the "soul" of the company, though, we also need to look at its board of directors, its backers.

Along with a purpose-driven business team, including our CEO Michael Gershenzon, Dr. Eugene Lipov and I were part of the team that launched Stella in 2020, three weeks before COVID made landfall. Stella was initially formed to make SGB available to everyone, not just Special Forces Operators. Our revenue entirely depended on a live procedure performed by a physician standing as close to a patient as a dentist. COVID threw a massive wrench in the plan.

This was a defining moment for our endeavor, and those who have backed this effort. Most backers would count the cost and throw in the towel in the face of social distancing, extensive testing protocols, and travel barriers. Stella's board met, talked and came back to us with this directive: "There has never been a time when effective treatments for trauma are more critical— we want you to get it right for the sake of everyone who is suffering."

In that initial year, we had time and space to get it right, resulting in an advanced SGB protocol called "DSR SGB" that is delivered over two days at a clinic with a success rate above 80% even when we treat severe symptoms of Post-Traumatic Stress. (If you want to see the data on several thousand patients, reach out and I'll share it). I had time to write a 35-page white paper on SGB (email me if you want a copy) and engage with numerous media outlets to bring new insights about trauma and healing.

Stella Mental Health is now much more than an "advanced SGB company"— offering other treatments like Ketamine treatments (IV and Spravato), Ketamine integration therapy, a "SOF Protocol" that combines DSR SGB with Ketamine infusions, and Transcranial Magnetic Stimulation (TMS), in addition to traditional medication management.

Many who suffer from trauma have found validation in the New York Times best-selling book "The Body Keeps the Score" by Bessel Van Der Kolk. This brilliant book created a leap forward by promoting the insight that trauma leaves its mark on the body, even when we're not consciously aware of it. This book also helped people understand the cumulative impact of trauma events, rather than the conventional model that trauma is caused by a single event that shatters one's sense of safety and trust.

At the same time, I've heard too many of my patients voice the belief that trauma makes its mark in a permanent, un-healable way: "at the end of the day, the body keeps the score" they might whisper. It's time to evolve beyond this belief. With the right treatments and the right support, trauma does not get to "run up the score" on the body. If the body keeps the score, then effective treatments are literal gamechangers. The "score" our body keeps includes how we heal too.

In other words, there is so much hope given the treatments that are available now. Feeling broken doesn't mean you are broken. Posttraumatic stress injuries can be seen on a brain scan, and they can be healed with innovative treatments. PTSD should no longer be seen as a permanent label, or even a chronic condition in many cases. And it's not just trauma—we now have much more effective treatments for depression, anxiety, and other life-altering conditions like obsessive-compulsive disorder.

Patients who have struggled with chronic suicidal thoughts are telling us that Spravato and intravenous Ketamine have completely eliminated these thoughts. Ketamine + integration therapy has been particularly powerful for addressing moral injuries. People who have felt the weight of severe depression for decades—who have tried everything with no relief, are finally getting relief with TMS. And DSR SGB continues to pull people out of tailspins and restore them back to calm and control in their bodies.

Along the way, I've had to develop an elastic identity by stretching, growing, and learning new ways of thinking. I've had to work through walls of resistance, in myself, and in countless conversations with others. For instance, as a person of faith, I've had to understand how a medicine like Ketamine, when used in a medically responsible way, is not inconsistent with faith. (If you want to read about my process in reaching this understanding, do a web search for "Ketamine is Not God" and "Psychology Today").

Stella Mental Health will continue to adapt and overcome the stagnation that has prolonged the suffering of too many people. There has been too much fallout already. Stella is one way that I decided to fall out and help build what you need. What works for Special Forces Operators is exactly what protectors and defenders of all varieties need access to. When enough of you demand access to what is now available, the suffering will meet its end.

If you'd like to learn more about innovative treatment options available today, please reach out (Doc@thinlineadvisory.com). I'm happy to share research articles, an academic review on safety and efficacy, large-scale outcome data, independent film footage, and firsthand accounts from veterans and first responders demonstrating how innovative treatment is changing—and saving—lives.

12

SOULFUL TECHNOLOGY

In the ever-expanding universe of "Ai-integrated apps," there's a spectrum – from apps without a soul, to apps that are full of soul. MAGNUS | ONE is full of soul - and here's the backstory on this initiative.

The legend of John Henry highlights the essence of the spectrum I just referenced. John Henry was to *building railroads* what Michael Phelps is to *swimming*— a railway worker who hammered steel by hand faster than anyone alive. When the railroad company brought in a new steam-powered drilling machine, John Henry took a stand for his crew, refusing to let the machine replace his men. He challenged the railroad company to a contest— man vs. machine.

Everyone from far and wide assembled at the foot of a mountain. When the whistle sounded, John Henry and the machine began to dig - side by side, tunneling through sheer rock. Knowing that his team's livelihood was at stake, John Henry swung his steel drill with purpose, driving himself harder than any steel worker ever had. He narrowly beat the machine, but celebrations were short lived. He was so exhausted that he collapsed and died on the railroad track, surrounded by the rubble from his superhuman feat. And the railroad company got behind steam powered drills in a big way despite his triumph. John Henry's story came to symbolize courage, professional pride, and the unconquerable human spirit even as the industrial age put many workers out of a job.

History is repeating itself right now.

We're living through something like the Industrial Revolution. Only this time, it's not machines replacing muscle. It's machines mimicking mind and emotion. And the speed of it all? It feels like we're being pushed into a race we never agreed to run.

There's a full-out scrum happening—companies fighting to become kings of this new AI world. Some of them are honest about what they want: to dominate the market, cut costs, and replace as many humans as possible. They talk about "efficiency" and "scalability," but what they're really doing is trying to build a world where paid employees are unnecessary.

Sometimes companies like this are so in love with their own technology that you'll see them light up when discussing their "capabilities." Yet, when the conversation turns to the nuances of human suffering, their interest dulls. These companies are like the railroad executives in John Henry's time— they're getting behind the steam powered drill with a level of tunnel vision that starts and ends with their profitability.

If you've ever been treated like a number instead of a person, you know how this kind of thinking lands. It's cold. It's disconnected. And if we're not careful, it's going to bleed into everything—our relationships, our families, even how we see ourselves.

But not everyone building AI-integrated apps is coming from that place. Some are people-first. Some are asking the right questions—questions like:

> *How can technology, to include AI, help humans connect better?*
>
> *How can technology help us rise instead of replacing us?*
>
> *How can technology help more people reach the kind of potential John Henry came to symbolize?*

For me, this is the invisible dividing line. It's not about the tech—it's about the hearts behind it. I've been pitched by countless companies who want to harvest human potential and subject matter expertise to "scale" pseudo-connection with bots. MAGNUS | ONE is different. MAGNUS | ONE asks the right questions and uses technology in the service of human potential.

Here's something else we don't talk about enough: the people training these AI systems matter. If they don't understand real human connection—if they don't know what trauma feels like, or what loyalty looks like, or how it feels when you carry the weight of responsibility for other people's lives—then what they put into the system will be empty. And what we get back will be the same.

“Garbage in, garbage out” isn’t just a statistics phrase. It’s a human truth that applies to AI.

And let me be even more direct: some of the folks trying to build this stuff are people who struggle to connect with other humans in deep, honest ways. If you can’t look another person in the eye and sit with their grief, their fear, or their truth, you’re going to build tech that avoids those things too.

And ultimately, here’s what we’ll end up with— AI that looks like “connection” on the surface—but doesn’t offer anything real underneath.

This scares me. Because what I see, in my conversations with protectors and defenders, is a world already starving for real connection. People are isolated, drifting, scrolling, numbing—and AI can either be something that pulls us farther apart or something that helps bridge those gaps.

We’re running fast—**way too fast**—in a direction that could leave us lonelier than before. And the loneliness I’m seeing out there? It’s already corrosive. Pseudo-connection eats at the soul. It builds walls around the heart and disables our capacity to sustain real relationships with other humans.

You know this better than most. You’ve lived what real connection looks like. You’ve trusted people with your life. You know what sacrifice, shared danger, and hard truth do to bond people together. You know that camaraderie is just love by another name - not romantic love but love that is deep and true. This kind of connection can't be faked—not by a human and certainly not by a machine.

So, here's what matters now:

We need to pay attention to the people building these systems.

> *Are they driven by heart?*
>
> *Are they grounded in soul-level understanding of what humans need?*
>
> *Do they actually care about connection—or just about getting ahead?*

We need builders who aren't afraid of real people.

Builders who know how to sit in the trenches of mental warfare with those who suffer.

Builders who understand what it means to serve, not to replace.

MAGNUS | ONE is a team of people that get it, who genuinely care about protectors and defenders. We are people who do this work regularly, at close range, in the trenches of mental warfare.

The Chairman of the Board, Dr. Mitch Javidi, shares a background in working within the Special Operations Community. Because of this, we are both grounded in the "Five SOF Truths," a set of guiding principles written in 1987 by Colonel John Collins and upheld as central values within the U.S. Special Operations Command (USSOCOM).

The first SOF Truth is this: **"Humans are more important than hardware."**

The Special Operations Community loves its technology and has some unbelievably advanced technical capabilities. Even so, greater value is placed on the human who is using the equipment. Mission success depends on the training, instinct, judgment, discernment, creativity, and elasticity of human operators.

The MAGNUS | ONE team is fueled by love and respect for protectors and defenders, with no exceptions. Dr. Javidi is a people-first visionary with the mind of an academic and the soul of a poet. He developed the MAGNUS OVEA Theory — a "theory of theories" that blends psychology, neuroscience, behavioral science, and social-cognitive principles into a unified framework for human performance and well-being. Rather than being narrowly focused on professional effectiveness, the MAGNUS OVEA theory informs the mental, emotional, physical, relational, and identity dimensions of reaching our potential. This theory forms the framework for our holistic approach to supporting leadership and peak performance.

What started as a collaboration has turned into a wonderful friendship. My friend Mitch is an expansively generous host, deeply trusted member of my Tribe, and consummate humanitarian. What you'll feel from him is this: *I'm with you and I'm for you.* Dr. Mitch Javidi is the heart of MAGNUS.

Our CEO, Jeff Kingsfield, is a 12-time Ironman and Kona World Championship finisher that leads with a servant heart. Jeff is fun, and he has the kind of marriage many of us aspire to - competing in triathlons with the love of his life, his wife Kelly Ann. Want to understand Jeff's heart? While at Georgia Tech, out of everyone who tried out, he was selected as the backup mascot "Buzz," the yellowjacket known for his energetic routines, including the pregame "Buzz Flip." Jeff has that kind of spirit— as the title of his autobiographical book suggests, he loves to Inspire.

I work closely with many people on the team, including Tiffany Andras and Brian Ellis. Tiffany is a mindfulness expert with a warrior spirit who has jumped out of a plane with her own parachute. And the thing she's most proud of: "Being a mother and the human being my son is and is becoming. It has been my greatest gift to get to help raise such a wonderful son and will always be the greatest source of pride in my heart."

When I asked Brian about what he's most proud of, he said this: "I'm proud that I finally learned to stop performing for the world and start listening to the part of me that was built for meaning. I'm proud that I've reached a time in life where purpose is measured by whom I help rise. The thing I carry with the deepest pride is that I've shifted from building a résumé to building a legacy. A legacy of giving. A legacy of developing others. A legacy of using whatever strength, wisdom, and scar tissue I've earned to make other people stronger."

And our team at MAGNUS | ONE has incredible partners. One of the best Sheriffs and leaders I've met is Steve Luce, who heads the Indiana Sheriff's Association. Steve is a visionary leader, who sees three moves ahead of everyone else. More than that, though, Steve has the heart of a wrestling coach (a former role before he was elected Sheriff) and he motivates all of us through the morning videos he sends after his weightlifting sessions. Steve is a beautiful soul, with a wonderful family, who are integral to his living legacy of service.

These are examples of some of the people behind this technology - their soul, their motivation, and their heart. And I contribute my insights on psychologically minded leadership, trauma, healing, close relationships, and the concept of elasticity.

As a result, MAGNUS | ONE is not another leadership app, it's a daily lifeline for people who carry heavy responsibility and still rise to meet the challenge. It turns science, story, and lived experience into short, powerful moments of reflection that rebuild purpose and elasticity from the inside out. Instead of teaching leaders what to do, it strengthens who they are, so they can face pressure without losing their clarity, compassion, or mission. This is an effort I'm proud to support – please let me know if you want to learn more.

13

A NEW KIND OF SUPPORT

I don't blame you for your reluctance to seek help. In a sea of experts far and wide, you continue to tell me that there's a very small pool of people that really "get it."

You're right about this. I've noticed this too, and here's why I think this happens.

To begin with, what you need isn't generally trained in graduate school programs. For example, licensed providers often have little to no training in addressing things like:

- military and first responder culture,
- barriers to trust for veterans and first responders,
- the difference between "PTSD" and cumulative stress,
- moral injuries and survivor guilt,
- the psychology of use of force,

- unique risk factors and covert indicators of suicide risk among protectors and defenders,
- the effect and process of healing from suicide loss,
- the impact of shift work and chronic sleep disruption,
- the ways relationships are impacted by deployments and first responder work (including secondary trauma),
- the complex challenge of leading a public safety agency,
- and the psychological and identity related impacts of retirement or military transition.

These are just a few examples of many more knowledge gaps.

Many people mistakenly think that licensed professionals can simply apply their general training to effectively assist protectors and defenders. The truth is that working with those who have a warrior spirit isn't the same as working with the general population. The culture itself creates a psychological "magnetic field" — a force that bends communication, trust, vulnerability, and disclosure in ways that outsiders almost never see.

Most mental health tools, including symptom severity and suicide risk assessments, are built for normal environments, and most clinicians are trained to use instruments that assume the compass needle will point North like it's supposed to.

But in public safety or military culture, the field is different:

- Loyalty and silence pull the needle in one direction.

- Stigma and fear of job consequences pull it in another.
- Control, rank, and authority are powerful undercurrents.
- Hidden moral injuries and perceptions of betrayal can distort the reading further.
- Self-inflicted moral injuries, so often the story behind the story, can throw a well-intended clinician off course.
- And identity, mission, belonging, and values so sacred that you would literally die for them— these things exert their own gravitational pull.

For most experts, the result is that their "compass" is chasing a false north; it gives readings that don't match the terrain. These Doctors may be skilled, and they often get great outcomes with other kinds of clients and patients, but they're not calibrated for this world.

A rare expert who *is* calibrated — a "Doc" who understands the culture, speaks its language, and sees its invisible forces — is like a compass that doesn't get thrown off. It points true north, even when everything around it is distorting the needle. These kinds of professionals are exceptionally uncommon. They do exist, even within the mental health teams in some agencies, but they are few and far between.

This is because this calibration process occurs over many years of being functionally embedded with protectors and defenders. Even if graduate schools attempted to teach them, many of the insights you need can't be learned in a classroom. Personally, I didn't learn these kinds of insights in any clinical setting. I learned them in the woods, with groups of Marines, in secluded settings with retired special operators, or in totally confidential, non-clinical conversations with countless warriors and first responders.

In other words, the insights of "Docs" come from decades of sharing a special and unique relationship of deep trust with protectors and defenders. In my experience, being deeply embedded in non-traditional settings may be the only way a person can become a translator between two languages that almost no one speaks. *Very few are truly bilingual when it comes to mental warfare and the warrior spirit.*

And it's not just a gap in knowledge - a lack of insights - that's the problem. There is also something deeper that can't be trained. This thing is as rare as that quality in you that I call "the warrior spirit." *The kind of people that can earn your trust have their own type of warrior spirit— different from yours, but similar enough that you'll recognize it instinctively.*

Maybe it's the sense of deeper purpose that creates this mutual recognition—that what each of us does is not a "job"—but a "calling." Or perhaps it's the alignment of unspoken but shared values. Maybe we meet your courage with our own form of courage—the kind that communicates, "you can tell me anything at all, and I won't be shocked, disgusted, horrified, or scared." In any case, when you drop in with a Doc, you'll find that opening up and speaking candidly about your inner world comes naturally and without fear.

This is *who* you need.

But the problem goes even deeper, into *what* you need. Because what I continue to hear is that the help you're offered now rarely feels built for you, and it feels more like a risk than a resource.

This is what you've been telling me you need:

- Access to trust-based relationships where you're truly safe (i.e., with no judgment or career consequences).
- Actionable insights and strategies that reflect your culture, values, and ethos for addressing a range of complicated challenges.
- Leaders you can count on and embedded peer support teams that know how to have your back.

The gap between what you have and what you need jumps out once we consider how the system is currently resourced. Let's take the example of any number of public safety agencies. Right now, "agency wellness" is contracted to two types of outside support services - "mental health treatment" and, more recently, "coaching."

Mental health treatment focuses on the assessment, diagnosis, and reduction of clinical symptoms. *But not all suffering or struggles can, or should, be diagnosed*. Many of you don't have diagnosable mental-health disorders— but still face issues that require culturally-informed psychological insights around things like:

- cumulative stress
- high stakes hiring decisions

- pivotal leadership dilemmas
- identity fractures
- traumatic grief (including brothers and sisters killed or lost to suicide)
- media misrepresentation
- moral injuries (including public character attacks)
- survivor guilt
- transitional challenges (new roles, being sidelined by an injury, retirement)
- moral exhaustion and loss of meaning
- hair trigger irritability and spillover anger
- relational breakdowns, divorce, confronting infidelity
- difficulties reintegrating after trauma
- systemic issues accessing resources

These aren't medical diagnoses, but the everyday pressures that naturally arise from your calling. A strictly clinical model - organized around a diagnosis with a focus on symptom reduction - is limited when it comes to addressing these challenges. Even if struggles like "hair-trigger irritability" and "moral exhaustion" wouldn't lead to an official diagnosis, you still feel reluctant to have them recognized or documented by a clinician.

The emergence of "coaching" as a "no-stigma/no medical notes" alternative to "mental health treatment" is a fairly recent development. You'll see coaching branded in various ways, for example: "resilience coaching," "leadership coaching," "executive coaching," "performance coaching," "wellness coaching," "communication coaching," "professional success coaching," "career coaching," "elite operator coaching," "strategic business coaching," "command staff coaching," "team coaching," "life coaching," and "transformational coaching."

As this range of names suggests, the coaching market has exploded in the past few years. There is a wide range of abilities among those who are certified as coaches. Some are really good, and some are terrible and completely unregulated.

And ultimately, "coaching" cannot sufficiently address what you need either. This is because a coach's scope of practice is much more limited than most people understand.

ICF (the International Coaching Federation) is the leading global non-profit organization that sets professional standards for coaching.

Based on ICF standards, a "coach" may only:

- Engage in non-directive conversations
- Elicit client-generated solutions
- Help clients set personal goals

Let's be clear and fair at the same time. There are excellent coaches doing valuable work. Coaching, when practiced well and within its ethical boundaries, can be enormously helpful for targeted goals like goal setting, accountability, and specific skill development.

But coaching is **not** a substitute for deep psychological expertise.

By the ethical standards of the coaching profession, coaches are not supposed to advise, interpret psychological patterns, or offer guidance grounded in psychological theory or assessment.

Consultation and advising rooted in doctoral-level psychology is fundamentally different work. It is informed by years of intensive training in human behavior, cognition, motivation, personality, group dynamics, and risk—along with the ability to recognize what *not* to touch and when deeper issues are being activated. Someone with extensive psychological training brings years of rigorous education, supervised clinical experience, and ethical accountability that simply cannot be replicated by a coaching credential. This depth is critical when it comes to accurate interpretation, sound judgment, and psychological safety.

Coaches are not substitutes for professionals with deep psychological expertise because:

- They are ethically constrained from advising or interpreting behavior through a psychological lens, while doctoral-level consultants are explicitly trained to do so.

- They are not equipped to identify psychological risk, complexity, or unintended consequences that can emerge even in non-clinical, high-stakes advisory work.
- They lack the clinical training and supervised experience required to assess, interpret, and safely intervene in complex psychological dynamics without risking oversimplification, misinterpretation, or injury.

Let's stop collapsing very different forms of expertise into the same bucket.

Consultation and advising informed by doctoral-level psychological training is a different category of work altogether. It's not about providing treatment; it's about bringing deep insight into human behavior, decision-making, identity, and risk—insight that comes from years of advanced study and disciplined application. That depth changes what you see, how you interpret it, and what guidance you can responsibly offer.

Coaching has a place. Psychological expertise changes the game. Treating them as interchangeable serves no one—and can actually do harm.

So, if what you need doesn't fall into the category of traditional mental health treatment, and coaching isn't sufficient to meet the true scope of challenges either, what do you need that is currently missing?

To answer this, I reflected on the countless private phone calls I've received from warriors and first responders over the past decades. I'm not being called for a diagnosis or a treatment plan, but something else. I couldn't find a precise name for it, so I asked my husband, the person who knows me best, what he would call the thing I do. His answer was quick: "You're like a *psychological wayfinder*—you draw from your expertise and years of experience as the Doc they trust to help them see their challenges in a totally new light. You help them find ways to navigate all kinds of really hard things that require psychological insight."

This is why, in 2025, I launched **THIN LINE ADVISORY** as a new category of support. Through THIN LINE ADVISORY, I offer highly specialized consultation and advising for organizations, departments and agencies that support veterans and first responders.

In this capacity, I provide translation, strategy, meaning, reconnection, and guidance at the deepest level of human experience, in a way that is attuned to your culture and values. This is a non-clinical service. As such, I do not diagnose, treat, or keep medical notes. Within this consultation role, I can respond to and resource a wide range of non-clinical needs. I am, as my husband observed, a *psychological wayfinder*.

Six applications of THIN LINE ADVISORY

1) Mentoring New Leaders Through High-Stakes Emotional Terrain

New leaders are thrust into sudden responsibility for people facing extreme stress, cumulative trauma, and life-or-death decision-making. I mentor new leaders around how to build trust, communicate effectively under pressure, and handle emotionally charged situations with clarity and confidence. I provide guidance that helps prevent leadership missteps that break trust and create perceived administrative betrayals within agencies. So often, it's not what leaders do, but what they don't know to do that breaks trust. For a clear example of this, you can read my Police1 article, "Administrative Betrayal: Why the Silence of Leaders is the Greatest Trauma for many LEOs."

2) **Advising Command Staff on High-Stakes Challenges that impact recruitment and retention**

Pivotal decisions create culture, and culture is what people embrace or walk away from.

Let's take suicide risk in a department as one example. Suicide within public safety communities is complex, deeply cultural, and often tied to unspoken norms, identity issues, betrayal traumas, or moral injury. The ways that leaders talk about risk, address risk within their agencies, and memorialize those who have died by suicide are defining moments for agency leadership.

To address this, I brief command staff on how distress actually presents in high-velocity, high-stigma professions, what warning signs are easily missed in protector/defender cultures, and how leaders can set a culture that addresses mental warfare directly without breaking trust. This is not standard suicide prevention training— it comes from years of being embedded in these communities.

Suicide risk is one of many critical areas of focus; other areas include collective grief after a line-of-duty death, ethical dilemmas for leaders, public character assassination (of a leader or agency as a whole), media misrepresentation, sensitive personnel issues, and many more.

3) **Equipping Peer Support Teams with Tools and Language They've Never Been Taught**

Peer support programs are among the most effective tools in public safety, yet they're often created and then left to operate without guidance or support. In many departments, peers end up serving as de facto substitutes for mental health clinicians despite having little to no psychological training.

I partner with peer teams to provide clinical consultation and support them as frontline emotional responders in their agencies. I help peers understand trauma, respond skillfully to disclosures of shame, guilt, or betrayal, support colleagues going through identity transitions (injury, retirement, forced leave), navigate conversations around moral injury and meaning, and avoid common pitfalls that accidentally shut people down. This results in a more confident, better-prepared peer force that can intervene earlier and more effectively.

Peers within veteran-focused organizations need the same guidance and support. One of the organizations I've been actively working with is the Veteran Mentor Project, Inc.

(VMPI) led by James and Luv Espinoza, both Army veterans. VMPI was founded in 2021 after James lost his brother, retired Army Lieutenant Colonel Ponce Espinoza, to suicide.

In honor of Ponce, James and Luv have transformed grief into a movement inspired by hope. They specifically saw the need for more connection and ongoing support for military veterans and veteran first responders in the transition back to civilian life. In response, James and Luv co-founded VMPI with the goal of mentoring America's heroes through transitions. Their team of veteran and first responder mentors are enlisted, officers, and first responders of all ranks who take their convictions and turn them into action. Reach out if you want to come on mission with us— it's an amazing initiative led by a team of mission-driven warriors.

Engaging me to train, consult, and collaborate with embedded peer teams is probably the most powerful and effective way to engage me as a "force multiplier" for a public safety agency, department, or veteran-focused organization. And I love this kind of work because partnering with those who have a warrior spirit (including the "Docs" I meet along the way) fulfills my deepest purpose.

4) **Bridging to Clinical Care with "Docs" on the Ground**

When a challenge requires clinical treatment, I help peer teams recognize this and connect their brothers and sisters with "Docs on the ground." Some agencies already have these providers; others need a bridge. The peer team and I serve in that role—strengthening inhouse programs or helping connect you with culturally informed, statelicensed Docs in your local community.

At times, you've shared that you don't trust the system or can't open up to providers within your agency. In those cases, I work with peer teams to identify alternative options and support meaningful, trustbased connections to care.

Through years of reflection and writing, including my book *WARRIOR*, I've come to understand the difference between "Doctors" and "Docs."

Doctors often rely on credentials and early assessments, positioning themselves as the experts in the room. Their work is defined as a job.

Docs see this work as a calling. They earn trust through listening, collaboration, and respect for the lived experience of warriors and first responders. They understand that trust takes time—and that there is often a story behind the story that only emerges when it is safe to do so.

Docs meet that moment with steadiness and humility, walking alongside people through their hardest seasons. Over time, they naturally work in collaboration, focused on one thing: getting it right for the people they serve.

If you're building or supporting a peer team, ask whether the providers you rely on are simply qualified—or truly called.

When the answer matters, I'm here to help you identify, connect with, and build bridges to Docs your people can trust.

5) **Helping Departments Build Trust After Critical Incidents**

Critical incidents are a high-stakes time for leaders and their teams. They are make-or-break moments for command staff, especially new leaders. This is where the values of a leader, or the culture of the organization, come into sharp focus. What a leader does or doesn't do can unintentionally damage their team and shorten the arc of their careers.

Traditional therapy focuses on treating the trauma exposure of an individual responder; it does not typically address organizational trust, cultural fallout, or group reintegration. I understand how these events fracture teams, shift relationships, and affect leadership credibility. I advise command staff on how to communicate, structure support, and address the human factors that influence long-term organizational health. One of my most requested resources is a 4-page asset I made for my clients that is titled, "10 Ways Well-Intentioned Leaders Damage Their Teams during a Critical Incident."

6) **Providing a safe outlet and strategic consultation for Public Safety Leaders**

The role of a public safety leader today is defined by constant strain and difficult, high-consequence choices that few outside the job can grasp. For example, police chiefs today carry an unrelenting burden driven by nonstop responsibility, constant

> crisis management and political pressures that undermine their ability to protect their officers.
>
> Chiefs are never "off duty": they are on call at all hours, living in a state of continuous hypervigilance. Every problem is funneled to them, from operational failures to personnel issues to community tensions. They become the center of accountability for events they often can't control, creating moral injuries when political or resource constraints prevent them from safeguarding their officers or departments.

A major source of stress comes from the weight of officer safety. Many who lead in public safety go to bed each night wondering whether their people had what they needed and whether everyone made it home safely. The phone becomes a permanent attachment, interrupting family life and eroding the boundary between personal and professional worlds. This intrusion strains relationships and reinforces the sense that the job always takes priority.

Additionally, chiefs shoulder the reputational risk of the entire profession; a single officer's misstep can impact the whole department and even the broader field of policing. This creates constant concern over public perception, political fallout and media scrutiny.

Despite these pressures, chiefs often feel they cannot seek traditional support like therapy. Mental health struggles may be used against them by political leaders or opponents who want to question their capability. Confidentiality worries and organizational dynamics create strong disincentives for accessing internal wellness programs. Even peer relationships within the command staff carry limits, as chiefs avoid burdening deputies who are already carrying their own stress. As so many have explained to me, "the idea of peer support between leaders at the same level is good in theory, but in reality the other leaders I would trust already have their plates more than full – I don't want to add anything else to their plates. I would rather carry the burden myself than offload it to someone who is already overloaded."

Because of these barriers, chiefs frequently operate without a safe outlet. Effective support then must come from credible, confidential, highly specialized advisors who exist completely outside the political and organizational chain. Without such support, the cumulative stress of leadership becomes a serious threat to their wellbeing and to the health of the departments they lead.

Military leaders have long benefited from this kind of support. Successful generals have private strategic advisors to help them see around corners, people who can help them anticipate future developments, potential challenges, and broader consequences. With THIN LINE ADVISORY, public safety leaders, peer teams, and organizations can now have the same.

My work with THIN LINE ADVISORY is deeply personal to me and grounded in partnership. I value thoughtful conversations with public safety leaders who genuinely care about their people and are seeking wellness approaches that go beyond check-the-box efforts. Much of this work is shaped by lived experience, cultural understanding, and time spent supporting individuals and leaders through grief and loss, moral injury, survivor guilt, and the complex decisions that influence trust, recruitment, and retention. If it feels helpful, I'd welcome the opportunity to connect and explore what support might look like—simply a conversation, at no cost. You're welcome to reach out anytime at Doc@thinlineadvisory.com.

14

VISION

The Right Pieces in the Right Place

For decades, we have tried to support veterans and first responders by rearranging the same incomplete pieces and hoping for a different outcome.

We ask peers to carry more than they should.
We assume therapy is the best natural starting point.
We expect leaders to hold everything with no safe outlet.
We offer suicide prevention that is based on civilian approaches.
We treat trauma as a permanent "disorder" that lives only in the mind.

When these efforts fall short, we often conclude that people "didn't engage," "weren't ready," or "resisted help." But what if the problem isn't the people?

What if the problem is that we've been operating under **false beliefs**—ideas that sound reasonable on the surface but quietly undermine everything we build?

One of the most persistent false beliefs is that more services automatically means better care. I think of this as the "menu of care" strategy. In practice, having a broad array of services is often overwhelming and poorly received by those whose trust in the system is low to begin with.

Another false belief is that any licensed clinician has the necessary qualities and skills to get good outcomes. A related false belief is that coaches are viable substitutes for people with deep psychological expertise (though some coaches are excellent in specific areas such as applying effective leadership principles or coaching around functional fitness goals). Another is that peer support is informal and therefore simple.

Perhaps the most damaging belief of all is that the order in which supports are offered doesn't really matter.

It does.

In high-risk, high-stigma professions, **sequencing can be the difference between engagement and withdrawal, between growth and shutdown, between life and loss**.

When someone with an injured fight-or-flight system is sent straight to therapy, we may be asking them to do something their body cannot yet support. When peers are not resourced with the right guidance, we unintentionally place them in harm's way. When leaders are given policies instead of perspective, culture erodes. When technology is used to replace human connection instead of extending it, trust dies straight out of the gate.

These gaps are not failures of effort. They are failures of design.

The hopeful truth is that **we do not need to start from scratch**. We already know what works—biological stabilization, culturally competent therapy, trusted peers, true experts who see what others don't, well-informed leaders, and accessible tools for growth. What has been missing is a way to **weave these elements together intentionally**, in the right order, with a clear understanding of how each prepares the ground for the next.

When we stop treating wellness as a collection of disconnected services and start treating it as an integrated system, something shifts. People engage earlier. Peers intervene with more confidence. Therapy goes deeper. Leaders make fewer missteps. Growth becomes continuous rather than crisis-driven.

This final chapter describes what is possible when we break from false beliefs and unhelpful approaches and build something new—not a program that looks good on paper, but a system that holds in real life. And it starts by getting the sequence right.

Veterans and first responders need systems that understand who they are, how injury actually unfolds, and how trust either accelerates healing—or blocks it completely. The most effective wellness programs are not defined by the number of services they offer, but by how those services are sequenced, integrated, and culturally aligned.

The unified model I envision is built around five core elements. These elements are not interchangeable, and they are not optional add-ons. Their order matters, because each one prepares the ground for the next.

When done well, this system becomes something rare in public safety and veteran care: a comprehensive ecosystem that adapts, responds, and grows with the people it serves.

Core Element I: Specialized Consultation — THIN LINE ADVISORY

Every effective system begins with orientation. Otherwise, veterans and first responders are lost in a sea of options. When trust is low to begin with and the mind and body feel overwhelmed by too many choices to "vet" there are two likely outcomes:

1) Get no support and continue to suffer
2) Pursue an option based on the personal stories of 1 or 2 peers (which often results in wasting money, seeking help from unsafe quacks, or trying risky and unproven approaches)

Before deciding what supports are needed, someone must help individuals and organizations understand:

- What kind of injury they're dealing with
- Where risk actually lives
- What supports are missing—or misaligned

- And what will unintentionally break trust if handled poorly

This is where THIN LINE ADVISORY sits—not as treatment, but as psychological wayfinding.

This work is intentionally non-clinical. There are no diagnoses, no treatment plans, no medical notes. Instead, it offers something that is often absent in high-stakes environments: translation, meaning, and strategic clarity.

This includes advising:

- Leaders facing ethical dilemmas, betrayal trauma, or public scrutiny
- Newly promoted command staff who are thrust into high stakes decisions
- Command staff facing suicide risk, line-of-duty deaths, or organizational fractures
- Peer teams who are absorbing far more emotional weight than they were ever trained for

Sequencing this correctly makes a difference.

Without this kind of "landscape view" advising:

- Organizations default to check-the-box solutions
- Clinical treatment is mismatched to the actual injury
- Peer teams are overwhelmed and under-resourced, and

- Leaders unintentionally damage trust during critical moments

THIN LINE ADVISORY helps organizations see clearly—often for the first time—what kind of support is actually required. It also provides peer teams with insights they cannot get anywhere else, because peers are often placed in emotionally complex situations without the psychological frameworks needed to navigate them safely.

In these ways, this first element sets the conditions for everything that follows.

Core Element II: Peer Support as a Front Door

I was once at a SOFWERX conference where the speaker shared that her group had researched who operators were most likely to talk with about their challenges. I'll never forget the results. Seared into my memory now is a slide that presented different kinds of people. And tenth on that list, after fellow operators, partners, friends, bartenders, barbers and several other categories, just above lawyers (who came in last) was "psychologists and other mental health professionals." Maybe that stings for some of us, but it's the truth.

For many veterans and first responders, peers—not clinicians—are the first safe point of contact. As I explained in my book *WARRIOR*, as healers, we may see ourselves as trustworthy based on our intentions and extensive training. But this doesn't mean that veterans and first responders do. The truth is that many of us must navigate a deep chasm of distrust before any real work can begin.

Peers are often trusted because they're known already. They understand the culture, the language, the gallows humor, the unspoken rules. When someone is struggling but not ready for therapy, peers are often the only door they are willing to walk through.

In this model, peer support is not treated as an afterthought or a volunteer add-on. Peers are recognized as what they truly are: frontline emotional responders.

They are often the first to notice:

- Subtle behavioral shifts
- Withdrawal or emotional constriction
- Loss of identity after injury, retirement or forced leave
- Shame, guilt, or moral conflict that doesn't yet look "clinical"

But peer support falters when peers are left alone.

Without guidance, peers can:

- Become de facto therapists without training
- Feel responsible for outcomes they cannot control
- Avoid hard conversations out of fear of "saying the wrong thing"

- Burn out under the cumulative emotional load of absorbing others' grief and trauma

Again, this is why sequencing matters.

Because THIN LINE ADVISORY comes first, peer teams are resourced, supported, and psychologically equipped. They learn:

- How trauma actually works in the nervous system
- Why their peers often conceal experiences of grief and loss
- How moral injury and betrayal show up differently than fear-based trauma
- How to respond to shame and survivor guilt without minimizing or trying to fix it
- How suicide risk can sit in a person's blind spot, and what questions can help us see true risk

Most importantly, peers are positioned as a bridge, not an endpoint. They accelerate trust and connection, then help guide their brothers and sisters toward the next layer of care—biological treatments (when needed) and Docs who get it.

Core Element III: Treatments That Heal Biological Injuries and Enhance Talk Therapy

Before people can process meaning, they must regain control inside their own bodies.

As I've written previously, repeated trauma exposure alters the nervous system. This is not a failure of coping. It is a biological injury.

Not everyone needs biological treatment, but many who have faced chronic traumatic environments can benefit. Others have hidden underlying conditions like depression, anxiety, or OCD that have never been addressed.

In this model, biological treatments are not viewed as stand-alone fixes, but as stabilizers that deepen and accelerate psychological healing. Dual Sympathetic Reset Stellate Ganglion Block (DSR SGB) is one example. By calming the sympathetic nervous system, DSR SGB restores internal control and creates the conditions necessary for therapy to work. It is the "primer to the paint" of good therapy.

But DSR SGB is not the only biological intervention that may be appropriate. The broader principle is this: When the body is calmer, the mind is more available. Other biological treatments are equally helpful for different kinds of challenges.

How easy is it to concentrate in therapy when you're stuck in a loop of suicidal thoughts? Intranasal Ketamine, an FDA approved treatment called Spravato may reduce or completely eliminate suicidal thinking in just a couple treatment sessions.

For those who are prescribed continuous regimens of anti-depressants, with limited relief, Transcranial Magnetic Stimulation (TMS) may be a game-changing procedure for resolving treatment resistant depression. And Ketamine infusions, when delivered in a medically responsible way, can be very helpful for addressing grief, moral injury, and experiences of betrayal and abandonment. It's likely that several additional promising modalities will become available as insurance-covered options in the next few years.

Offering biological treatments before talk therapy:

- Reduces crisis-level arousal
- Expands the window of tolerance
- Improves sleep and emotional regulation and
- Allows people to focus and stay present in therapy

Sequencing matters here too, not just in respect to therapy, but with the elements that lay the prior groundwork. When biological stabilization occurs after strategic wayfinding and peer connection:

- Treatments are better matched to individuals
- Treatment is framed in a no-stigma, injury-based way
- Treatment is integrated with—not separated from—therapy

This prevents the false split between "medical" and "psychological" care.

Core Element IV. Mental Health Treatment With "Docs" Who Truly Get It

Therapy is essential—but only when delivered by the right people, in the right way, at the right time.

Veterans and first responders do not just need trauma specialists. They need Docs—clinicians who understand protector culture, moral injury, survivor guilt, identity loss, betrayal trauma, and the difference between intrusive thoughts and intent.

A Doc understands that self-destructive thoughts are a common experience, not the challenge of an outlying individual. Personally, I've observed that these thoughts are so common that I've come to believe this is part of the path. Here's how I make sense of this.

During medical school, there's something that happens to many students during their second year. "Second-year medical student phenomenon" refers to a common experience where medical students, after learning about diseases in detail, begin to notice symptoms of those diseases in themselves.

As knowledge increases, awareness and pattern recognition increase, and normal bodily sensations or thoughts get interpreted through a new lens. This can lead to temporary worry or misattribution, even though nothing is actually wrong. This is the mind doing what it's designed to do: scan for patterns and make meaning based on recent exposure.

Similarly, in trauma-exposed populations, when people repeatedly witness death, injury, or suicide, their minds naturally become more familiar with those pathways. Thoughts about death may appear not because someone wants to die, but because their brain has been repeatedly exposed to it.

When training a group of police chiefs, I put it like this, "Here's a powerful way to change the culture— tell your people this: you'll see some really messed up things and because of this, your mind is going to go to dark places sometimes. That's not weakness—that's exposure. Having a dark or even self-destructive thought doesn't mean you're going to act on it; it means you've seen what we see. Those thoughts are common and predictable for people who see death and trauma. The danger isn't having the thought—the danger is carrying it alone. When you talk it through with someone you trust, it loses its grip."

After the training, a Chief told me that my message—that regularly seeing death and trauma leads to dark thoughts—was completely liberating. As he explained, "until you said that I didn't realize how heavy it was to feel like I'm not supposed to be impacted by all the messed-up stuff I've seen. That mentality starts early for us, as early as the training we get in the academy." He suggested bringing my training into public safety academies, saying "what you shared today, your whole take on suicide, how we need to think in a totally different way, set a new cultural understanding from the very start, look for different signs of risk, and take action in a completely different way, could change everything about how we enter the profession and deal with the stuff that comes up, when it happens."

So, when a warrior or first responder discloses dark and self-destructive thoughts, a Doc understands that this is a predictable form of mental warfare that is commonly experienced within protectors and defenders. Without fear or discomfort, a Doc engages in a frank and open discussion, putting this into the context I just shared, without any shame, which immediately lowers actual risk. And a Doc knows that suicidal thoughts in this population often exist on a continuum and must be handled with precision, not panic.

A Doc also understands that trauma is not the only wound. Many of the deepest injuries carried by those who serve are moral, not fear-based.

Moral injury arises when individuals:

- Survive when others didn't
- Feel responsible for outcomes they couldn't control
- Participate in, witness, or fail to prevent acts that violate sacred values
- Are exposed to attacks on their character or integrity, especially in a public forum

These injuries are often accompanied by profound shame, guilt, and loss of meaning—and they do not always respond well to traditional trauma-focused approaches. And as I've said, trauma exposure as the cause of suicide may be a "red herring." In fact, what I've repeatedly observed is that the even deeper, more invisible wounds caused by moral injuries, complicated grief, identity fracture, and perceived betrayal are often more lethal than exposure to scenes of trauma. This is where group-based modalities become especially powerful.

During my time within the Department of Veterans Affairs, I developed and ran a highly effective moral injury group which taught me that moral injury is best treated in a communal way. Other VA clinicians, such as Dr. Deirdre Lopez, have also developed impactful group-based approaches that allow participants to:

- Break isolation
- Normalize moral pain
- Restore meaning and values
- Rebuild identity in community

In this unified vision, therapy includes:

Individual work with trusted Docs

Group formats for moral injury and survivor guilt

Ongoing consultation and training among Docs themselves

Just as a warrior or first responder never stops training, a Doc never stops learning. They learn from other Docs, mutually expand collective cultural competence, and communicate this to the larger healthcare ecosystem.

Core Element V: MAGNUS | ONE — Continuous Access to Technology With Soul

Even the best human systems have limits.

- Peers are not always available.
- Therapy happens once or twice a week.

Growth does not occur only in appointments. As I wrote about in my book *WARRIOR*, much of the transformative work in therapy depends on what people do in the space between sessions.

MAGNUS | ONE exists to bridge these gaps and optimize growth between human encounters.

This is not generic mental health technology. It is technology with soul, built by people who understand service culture and psychological injury.

MAGNUS | ONE provides:

- 24/7 access to trusted resources
- Exercises that help users apply insights in real life
- A range of tools that promote continual, incremental growth
- Short, digestible videos that speak the language of the community

Its design matters.

Veterans and first responders are more likely to engage with brief, on-point content, practical tools, and resources that respect their time and intelligence.

MAGNUS | ONE becomes the connective tissue, reinforcing every other core element:

- It gives people shared language and tools
- It helps people integrate biological treatments
- It helps people take action between therapy sessions
- It does not replace human connection. It extends it.

Why Sequencing and Integration Matter

In this system, each element prepares the ground for the next.

- Informed advising ("wayfinding") clarifies needs and prevents missteps
- Peers build trust, open doors, and bridge to other forms of support and care
- Biological treatments restore control and set the optimal conditions for talk therapy
- Therapy addresses underlying psychological conditions, trauma, moral injury, and meaning
- Technology offers easy, personalized access to insights, exercises, and resources to sustain growth

Each element enhances the others.

Remove wayfinding, and everything becomes reactive.

Remove peers, trust collapses and healing never happens.

Remove biological treatments, and progress in therapy is limited.

Remove culturally competent Docs, and protectors drop out of care.

Remove technological capabilities, and growth fades between touchpoints.

Together, these five core elements form a program that is far greater than the sum of its parts.

Veterans and first responders do not need to be "fixed." They need to be understood, supported, and resourced intelligently.

This model honors their biology, culture, identity, moral pain, and capacity for growth. It does not promise that service will leave no marks. But it restores control, preserves dignity, and supports continual positive growth—one decision, one connection, one moment at a time.

This is my vision for what it looks like to build a structure of support that actually holds.

CONCLUSION

A New Set of Truths

Throughout these chapters, we've dismantled myths that have long held veterans and first responders back: that suicide is always preventable, that love alone protects someone from risk, that courage is lacking if help isn't sought, and that post-traumatic stress is a life sentence. These myths have done real damage, leaving our protectors isolated, burdened with shame, and unsure where to turn.

In this book, I've presented a different set of truths. Some of these truths are reminders to us. For example, first responders are courageous beyond measure, trained to operate under extreme pressure, and capable of compartmentalizing pain that would overwhelm most civilians.

Some of these truths are new. For example, their resilience, paradoxically, becomes an Achilles heel: they hide distress, remain operational even while on the ropes with the voice of despair, and often never show warning signs when they're struggling. Even though they are so good at reading other people, they may not see the suicide risk that is sitting in their blind spot.

Mental warfare is not uncommon— it is an unavoidable part of pursuing the call to serve. Recovery, wellness, and suicide prevention are not about courage, character, or morality— they are about **safety, trust, and access to the right interventions.**

The Path Forward

The path forward is clear: reject false beliefs, embrace forward-thinking policies, and create safe spaces for veterans and first responders to get help without fear. Recovery is possible. Lives can be changed and saved. Systems can evolve to support those who have long sacrificed for others.

Those who risk it all to protect us deserve support that is both effective and culturally informed. If we fail to change, they will continue to suffer. So, let's fall out, together, and build this.

ACKNOWLEDGEMENTS

Writing a book is never a solitary act. My work is influenced by a broader community whose insights, dedication, and generosity contribute significantly to its development. This book is no exception. It exists because of a constellation of relationships—built over time, sustained by trust, and grounded in a shared commitment to those who protect and defend our communities and our nation.

First and foremost, this book would not exist without the unique and sacred relationships I have been privileged to share with first responders and veterans over the past two decades. Every insight in these pages is rooted in what you have entrusted to me. Each story—especially those shared at great personal cost—has helped illuminate not only the depth of pain carried by protectors and defenders, but also the extraordinary strength, loyalty, and moral seriousness with which that pain is borne. Every individual who trusted me with their most challenging, private, and often unspeakable experiences contributed to a clearer understanding of what meaningful support truly requires. Your honesty has shaped my thinking, refined my questions, and continually called me to do better. I hold your trust as sacred.

Over the years of collaboration, it has become evident that providing effective support requires a reassessment of the existing models and frameworks through which assistance is delivered. Support must be culturally fluent, operationally realistic, and grounded in respect for identity, mission, and values. The light that shines through this book is not mine alone—it reflects the collective wisdom of protectors and defenders who were willing to let me see what they do not share with others, and what many systems still fail to recognize.

Each book I write also carries the fingerprints of people who step forward at exactly the right moment—those who read early drafts, challenge my assumptions, strengthen my arguments, and remind me why the work matters when the process feels heavy. In selecting a foreword writer, it was clear that your expertise and character would be essential. Your demonstrated courage in questioning established perspectives and your credibility allow readers to gain a broader understanding of wellness and the unique needs of protectors and defenders. CZ, your moral integrity, practical experience, and steadfast dedication to the Tribe of protectors and defenders are greatly valued. I am deeply honored by your support and friendship.

I am also profoundly thankful for my circle of confidential and trusted pre-release reviewers—the majority of whom belong to both tribes, having served as first responders and as veterans, and who have spent meaningful time in both worlds. Your willingness to read closely, speak candidly, and protect the integrity of this work was an extraordinary gift. Thank you to Chris, Tony, Ted,

James, Dave, Beau, Jeff, Mary, Paul, Daniel, Erin, Steve, Mitch, Jeff, Brian, Tiffany, Jonny, Gail, Ryan, and Brandon. Your feedback, encouragement, and careful attention ensured that these pages remained grounded in reality and respectful of the communities they are meant to serve.

Finally, to everyone who has walked alongside me in this work—often quietly, often without recognition—please know that your presence matters more than you may realize. This book is a reflection of shared effort, shared courage, and shared responsibility. It stands as a testament to what becomes possible when we listen deeply, honor the wisdom of lived experience and commit ourselves to supporting protectors and defenders in ways that truly meet the need.

With gratitude and respect,

Doc Springer

REFERENCES AND FURTHER READING

Begemann, M. J. H., Schutte, M. J. L., et al. (2023). *Childhood trauma is associated with reduced frontal gray matter volume: A large transdiagnostic structural MRI study. Psychological Medicine, 53*(3), 741–749. https://doi.org/10.1017/S0033291721002087

Belen, V. (2025, January 25). *LA wildfires stress firefighters, highlighting mental health needs. Thin Line News.* https://thinlinenews.com/la-wildfires-stress-firefighters-highlighting-mental-health-needs/

Bennett, A. (2020, May 18). *Retreats help Marines protect each other from suicide. Connecting Vets.* https://www.audacy.com/connectingvets/articles/retreats-help-marines-protect-each-other-from-suicide

Bérubé, A., Turgeon, J., et al. (2023). *Emotion recognition in adults with a history of childhood maltreatment: A systematic review. Trauma, Violence, & Abuse, 24*(1), 278–294. https://doi.org/10.1177/15248380211029403

Blosnich, J. R., Dichter, M. E., Cerulli, C., Batten, S. V., & Bossarte, R. M. (2014). Disparities in adverse childhood experiences among individuals with a history of military service. *JAMA psychiatry*, *71*(9), 1041–1048. https://doi.org/10.1001/jamapsychiatry.2014.724

Bremner, J. D. (2003). *Long-term effects of childhood abuse on brain and neurobiology. Child and Adolescent Psychiatric Clinics of North America, 12*(2), 271–292. https://doi.org/10.1016/s1056-4993(02)00098-6

Carr, J. (2022). *Overworked California firefighters struggle with PTSD, suicide, fatigue, intensifying wildfires. CalMatters.* https://www.firefighterclosecalls.com/overworked-firefighters-struggle-with-ptsd-suicide-fatigue-intensifying-wildfires/

Coppersmith, D. D. L., Jaroszewski, A. C., Gershman, S. J., Cha, C. B., Millner, A. J., Fortgang, R. G., Kleiman, E. M., & Nock, M. K. (2024). Do people know how suicidal they will be? Understanding suicidal prospection. *Suicide & life-threatening behavior*, *54*(4), 750–761. https://doi.org/10.1111/sltb.13087

Cross, D., Negar, F., et al. (2017). *Neurobiological development in the context of childhood trauma. Clinical Psychology (New York), 24*(2), 111–124. https://doi.org/10.1111/cpsp.12198

Dalgleish, T., Moradi, A. R., et al. (2001). *An experimental investigation of hypervigilance for threat in children and adolescents with post-traumatic stress disorder. Psychological Medicine, 31*(3), 541–547. https://doi.org/10.1017/S0033291701003567

Javidi, M., Kingsfield, J., & Springer, S. (2025). *Elastic Identity. Readiness Network, Inc.*

Javidi, M., Kingsfield, J., & Springer, S. (2025). *Elasticity of identity theory: The science behind adaptation, performance, and thriving*. LinkedIn. https://www.linkedin.com/pulse/elasticity-identity-theory-dr-mitch-javidi-lmxze/

Javidi, M. (2025). *Beyond bouncing back: Why elasticity is the new gold standard for human growth*. LinkedIn. https://www.linkedin.com/pulse/beyond-bouncing-back-why-elasticity-new-gold-standard-javidi-zytje/

Katon, J. G., Lehavot, K., Simpson, T. L., Williams, E. C., Barnett, S. B., Grossbard, J. R., Schure, M. B., Gray, K. E., & Reiber, G. E. (2015). Adverse Childhood Experiences, Military Service, and Adult Health. American journal of preventive medicine, 49(4), 573–582. https://doi.org/10.1016/j.amepre.2015.03.020

Khan, A. J., Campbell-Sills, L., Sun, X., Kessler, R. C., Adler, A. B., Jain, S., Ursano, R. J., & Stein, M. B. (2021). Association Between Responsibility for the Death of Others and Postdeployment Mental Health and Functioning in US Soldiers. *JAMA network open*, *4*(11), e2130810. https://doi.org/10.1001/jamanetworkopen.2021.30810

Levy, M. S. (1998). *A helpful way to conceptualize and understand reenactments. Journal of Psychotherapy Practice and Research, 7*(3), 227–235. https://www.ncbi.nlm.nih.gov/pmc/articles/PMC3330499/

Lipov, E. (2023, May 11). *Survey reveals that renaming post-traumatic stress 'disorder' to 'injury' would reduce stigma. Cureus, 15*(5), e38888. https://doi.org/10.7759/cureus.38888

Lipov, E., & Candido, K. (2021). The Successful Use of Left-Sided Stellate Ganglion Block in Patients That Fail to Respond to Right-Sided Stellate Ganglion Block for the Treatment of Post-Traumatic Stress Disorder Symptoms: A Retrospective Analysis of 205 Patients. *Military medicine, 186*(11-12), 319–320. https://doi.org/10.1093/milmed/usab150

Lipov, E. G., Jacobs, R., Springer, S., Candido, K. D., & Knezevic, N. N. (2022). Utility of Cervical Sympathetic Block in Treating Post-Traumatic Stress Disorder in Multiple Cohorts: A Retrospective Analysis. *Pain physician, 25*(1), 77–85.

Lipov, E. G., Joshi, J. R., Lipov, S., Sanders, S. E., & Siroko, M. K. (2008). Cervical sympathetic blockade in a patient with post-traumatic stress disorder: a case report. *Annals of clinical psychiatry: official journal of the American Academy of Clinical Psychiatrists, 20*(4), 227–228. https://doi.org/10.1080/10401230802435518

Lipov, E. G., & Ritchie, E. C. (2015). *A review of the use of stellate ganglion block in the treatment of PTSD. Current Psychiatry Reports, 17*(8), Article 599. https://doi.org/10.1007/s11920-015-0599-4

Lupien, S. J., OuelletMorin, I., et al. (2016). *Epigenetics and neuroendocrinology*. In *From vulnerability to neurotoxicity: A developmental approach to the effects of stress on the brain and behavior* (pp. 3–48). Springer International Publishing.

Lynch, J. H., Mulvaney, S. W., Kim, E. H., de Leeuw, J. B., Schroeder, M. J., & Kane, S. F. (2016). Effect of Stellate Ganglion Block on Specific Symptom Clusters for Treatment of Post-Traumatic Stress Disorder. *Military medicine, 181*(9), 1135–1141. https://doi.org/10.7205/MILMED-D-15-00518

Lynch, J. H., Muench, P. D., Okiishi, J. C., Means, G. E., & Mulvaney, S. W. (2021). *Behavioral health clinicians endorse stellate ganglion block as a valuable intervention in the treatment of traumarelated disorders. Journal of Investigative Medicine, 69*(5), 989 -993. https://doi.org/10.1136/jim-2020-001642

Maguen, S., Griffin, B. J., Vogt, D., Hoffmire, C. A., Blosnich, J. R., Bernhard, P. A., Akhtar, F. Z., Cypel, Y. S., & Schneiderman, A. I. (2023). Moral injury and peri- and post-military suicide attempts among post-9/11 veterans. *Psychological medicine, 53*(7), 3200–3209. https://doi.org/10.1017/S0033291721005274

McDonald, A. R., Prost, S. G., et al. (2022). *The contribution of adverse childhood experiences to workrelated stressors among law enforcement officers. Policing: A Journal of Policy and Practice, 16*(1), 44–62. https://doi.org/10.1093/police/paab047

McLean B. (2015). Safety and Patient Acceptability of Stellate Ganglion Blockade as a Treatment Adjunct for Combat-Related Post-Traumatic Stress Disorder: A Quality Assurance

Initiative. *Cureus*, *7*(9), e320. https://doi.org/10.7759/cureus.320

Mulvaney, S. W., Lynch, J. H., Hickey, M. J., Rahman-Rawlins, T., Schroeder, M., Kane, S., & Lipov, E. (2014). Stellate ganglion block used to treat symptoms associated with combat-related post-traumatic stress disorder: a case series of 166 patients. *Military medicine*, *179*(10), 1133–1140. https://doi.org/10.7205/MILMED-D-14-00151

Mulvaney, S. W., Mahadevan, S., Desronvilles, R. J., Jr, Dineen, K. J., & Rae Olmsted, K. L. (2025). Three-Month Durability of Bilateral Two-Level Stellate Ganglion Blocks for Traumatic Brain Injury: A Retrospective Analysis. Biomedicines, 13(7), 1526. https://doi.org/10.3390/biomedicines13071526

Rae Olmsted, K. L., Bartoszek, M., Mulvaney, S., McLean, B., Turabi, A., Young, R., Kim, E., Vandermaas-Peeler, R., Morgan, J. K., Constantinescu, O., Kane, S., Nguyen, C., Hirsch, S., Munoz, B., Wallace, D., Croxford, J., Lynch, J. H., White, R., & Walters, B. B. (2020). Effect of Stellate Ganglion Block Treatment on Posttraumatic Stress Disorder Symptoms: A Randomized Clinical Trial. *JAMA psychiatry*, *77*(2), 130–138. https://doi.org/10.1001/jamapsychiatry.2019.3474

O'Brien, P. and Campbell, D.G. (2021) Wildland Firefighter Psychological and Behavioral Health: Preliminary Data from a National Sample of Current and Former Wildland Firefighters in the United States. Sourced online 9/13/25. https://link.springer.com/article/10.1186/s42408-026-00449-9

Pressfield, S. (1998). *Gates of fire: An epic novel of the battle of Thermopylae*. Doubleday.

Roncoroni, J., & Springer, S. (2019). *Beyond the military: A leader's handbook for warrior reintegration*. Lioncrest Publishing.

Roth, S. L., Andrews, K., Protopopescu, A., Lloyd, C., O'Connor, C., Losier, B. J., Lanius, R. A., & McKinnon, M. C. (2022). Mental health symptoms in Public Safety Personnel: Examining the effects of adverse childhood experiences and moral injury. *Child abuse & neglect*, *123*, 105394. https://doi.org/10.1016/j.chiabu.2021.105394

Ruderman White Paper on Mental Health and Suicide of First Responders (2016). Sourced online 9/13/25. Ruderman White Paper on Mental Health and Suicide of First Responders – Ruderman Family Foundation (Wednesday, 9/24/25). https://rudermanfoundation.org/white_papers/the-ruderman-white-paper-update-on-mental-health-and-suicide-of-first-responders/

Springer, S. (2016, September 23). *Can a Harvard Degree Actually Reduce Patient Trust? Re-thinking the public display of a diploma. Psychology Today*. https://www.psychologytoday.com/us/blog/free-range-psychology/201609/can-harvard-degree-actually-reduce-patient-trust

Springer, S. (2016, October 15). *Reflections on veteran suicide, veteran resilience and tribe. Psychology Today*. https://www.

psychologytoday.com/us/blog/free-range-psychology/201610/reflections-veteran-suicide-veteran-resilience-and-tribe

Springer, S. (2016, November 14). *Crisis Response Models for Suicide Are Not Enough: Envisioning a better, stronger model—a communal response to crisis. Psychology Today.* https://www.psychologytoday.com/us/blog/free-range-psychology/201611/crisis-response-models-for-suicide-are-not-enough

Springer, S. (2018, April 25). *The power of tribe. Psychology Today*. https://www.psychologytoday.com/za/blog/free-range-psychology/201804/the-power-of-tribe

Springer, S. (2018, June 11). *Is suicide selfish? Psychology Today*. https://www.psychologytoday.com/us/blog/free-range-psychology/201806/is-suicide-selfish

Springer, S. (2018, August 29) Suicide Risk Among Society's Emotional First Responders: Protecting those in ministry roles who serve as "lighthouses" to many. Psychology Today. https://www.psychologytoday.com/us/blog/free-range-psychology/201808/suicide-risk-among-societys-emotional-first-responders

Springer, S. (2018, October 25). *The Reason to Go On After Serving in the Military: With the right support, Warriors can walk the wasteland of transition. Psychology Today.* https://www.psychologytoday.com/us/blog/free-range-psychology/201810/the-reason-go-after-serving-in-the-military

Springer, S. H. (2019, March 20). Career transitions for NFL players and combat warfighters. Psychology Today. https://www.psychologytoday.com/au/blog/free-range-psychology/201903/career-transitions-for-nfl-players-and-combat-warfighters

Springer, S. (2019, March 27). *A tactical analysis of the war on hopelessness. Psychology Today*. https://www.psychologytoday.com/us/blog/free-range-psychology/201903/a-tactical-analysis-of-the-war-on-hopelessness

Springer, S., & Ergo, M. (2019, June 14). *Calling suicide shameful is the wrong call. Marine Corps Times*. https://www.armytimes.com/newsletters/daily-news-roundup/2019/06/14/calling-suicide-shameful-is-the-wrong-call/

Springer, S. (2019, August 2). *Why your relationship is at risk when you leave the military. Army Times*. https://www.armytimes.com/education-transition/2019/08/02/opinion-why-your-relationship-is-at-risk-when-you-leave-the-military/

Springer, S. (2019, August 14). *New promise for treating trauma: An injection may give people a "second shot at a happy life." Psychology Today*. https://www.psychologytoday.com/us/blog/free-range-psychology/201908/new-promise-treating-trauma

Springer. S. (2019, November 15) Why Marines Make Exceptional Social Workers: Profiling some of our best mental warfare warfighters. Psychology Today. https://www.psychologytoday.

com/us/blog/free-range-psychology/201911/why-marines-make-exceptional-social-workers

Springer, S. (2020, April 1). *An invisible enemy stalks us—firearm suicides, and how you can fight it* [Opinion]. *Army Times*. https://www.armytimes.com/opinion/2020/04/01/the-invisible-enemy-that-stalks-us-firearm-suicides-and-how-you-can-fight-it/

Springer, S. (2020, April 19). *An open letter to our nation's warriors during these trying times* [Opinion]. *Military Times*. https://www.militarytimes.com/opinion/commentary/2020/04/19/an-open-letter-to-our-nations-warriors-during-these-trying-times/

Springer, S. (2020, June 13). *How to support those who protect us*. *Psychology Today United Kingdom*. https://www.psychologytoday.com/gb/blog/free-range-psychology/202006/how-to-support-those-who-protect-us

Springer, S. (2020, September 18). *Lessons learned from a patient suicide attempt*. *Psychology Today*. https://www.psychologytoday.com/us/blog/free-range-psychology/202009/lessons-learned-patient-suicide-attempt

Springer, S. (2020, October 20). *Five Common Myths About PTSD: PTSD stigma is fueled by a misunderstanding of emotional trauma. Psychology Today.* https://www.psychologytoday.com/us/blog/free-range-psychology/202010/5-common-myths-about-ptsd

Springer. S. (2021, January 13). *The Most Promising PTSD Treatment You've Never Heard Of: Innovating by fusing biological and psychological approaches. Psychology Today.* https://www.psychologytoday.com/us/blog/free-range-psychology/202101/the-most-promising-ptsd-treatment-youve-never-heard

Springer, S. (2021, March 10). *Let's rethink the relationship between mental health healers and those they serve. Military.com.* https://www.military.com/daily-news/opinions/2021/03/10/lets-rethink-relationship-between-mental-health-healers-and-those-they-serve.html

Springer. S. (2021, September 9). *Two Types of Providers: "Doctors" and "Docs". Becoming a "Doc" is critical for helping veterans heal. Psychology Today.* https://www.psychologytoday.com/us/blog/free-range-psychology/202109/two-types-providers-doctors-and-docs

Springer, S. (2022, May 4). *The Human Toll of Trauma for Law Enforcement Officers. What happens to those whose job requires them to see us at our worst? Psychology Today.* https://www.psychologytoday.com/us/blog/free-range-psychology/202205/the-human-toll-trauma-law-enforcement-officers

Springer, S. (2023, February 16). *Bravery Does Not Predict Good Outcomes in Therapy: Why courage is more important than bravery for mental wellness. Psychology Today.* https://www.psychologytoday.com/us/blog/free-range-psychology/202302/bravery-does-not-predict-good-outcomes-in-therapy.

Springer, S. (2024, February 14). *Administrative betrayal: Why the silence of leaders is the greatest trauma for many LEOs. Police1.* https://www.police1.com/what-cops-want/administrative-betrayal-why-the-silence-of-leaders-is-the-greatest-trauma-for-many-leos

Springer, S. (2024, September 6). *Productive and Unproductive Suffering. Three critical features distinguish suffering that can help people heal. Psychology Today. https://www.psychologytoday.com/us/blog/free-range-psychology/202409/productive-and-unproductive-suffering*

Springer, S. (2024, September 13). *What people really mean when they say 'suicide is selfish'. Police1.* https://www.police1.com/health-wellness/what-people-really-mean-when-they-say-suicide-is-selfish (Police1)

Springer, S. (2024, October 7). *Suicide risk assessment: A critique. Psychology Today*. https://www.psychologytoday.com/us/blog/free-range-psychology/202410/suicide-risk-assessment-a-critique

Springer, S., & Roncoroni, J. (2019, November 15). *How the stress of military transition leads to regrettable career and relationship decisions. Military Times.* https://www.militarytimes.com/opinion/commentary/2019/11/15/how-the-stress-of-military-transition-leads-to-regrettable-career-and-relationship-decisions/

Springer, S. (2021). *WARRIOR: How to support those who protect us*. Hidden Ivy Press.

Springer, S., & Sugrue, M. (2022). *Relentless Courage: Winning the battle against frontline trauma*. Hidden Ivy Press.

Springer, S., Whitmer, P., Steinlin, M., Gray, L., & Blankfield, J. (2024). Optimizing clinical outcomes with stellate ganglion block and trauma-informed care: A review article. *NeuroRehabilitation, 55*(3), 385–396.

Springer, S. (2025). *Should we rethink the concept of "resilience"? LinkedIn*. Retrieved February 1, 2026, from https://www.linkedin.com

Springer, S. (2025, March 22). *The unseen toll of leadership: Chiefs speak out on the human cost of command. Police1*. https://www.police1.com/wellness-week/the-unseen-toll-of-leadership-chiefs-speak-out-on-the-human-cost-of-command/(police1.com)

Springer, S. (2025, May 14). *Fight or flight in overdrive. Police Chief Magazine*. https://www.policechiefmagazine.org/fight-flight-overdrive/ (policechiefmagazine.org)

Springer, S. (2025, May). *A new treatment brings new hope for law enforcement officers: Migrating what works in Special Forces to the LEO community. FOP Journal*. National Fraternal Order of Police.

Springer, S. (2025, December 3). *Childhood trauma and public safety careers: Is there a connection? FireRescue1*. https://www.firerescue1.com/fire-careers/childhood-trauma-and-public-safety-careers-is-there-a-connection/

Springer, S. (2026, January 28). *Why suicide risk in first responders hides in plain sight. CRACKYL Magazine*. https://www.crackyl.com/why-suicide-risk-in-first-responders-hides-in-plain-sight/

Summers, M. R., & Nevin, R. L. (2017). Stellate Ganglion Block in the Treatment of Post-traumatic Stress Disorder: A Review of Historical and Recent Literature. *Pain practice: the official journal of World Institute of Pain*, *17*(4), 546–553. https://doi.org/10.1111/papr.12503

Tedeschi, R. G., & Calhoun, L. G. (1996). The Posttraumatic Growth Inventory: measuring the positive legacy of trauma. *Journal of traumatic stress*, *9*(3), 455–471. https://doi.org/10.1007/BF02103658

U.S. Department of Veterans Affairs. (2022). *2022 National Veteran Suicide Prevention Annual Report*. Office of Mental Health and Suicide Prevention. https://www.mentalhealth.va.gov/docs/data-sheets/2022/2022-National-Veteran-Suicide-Prevention-Annual-Report-FINAL.pdf

Van Dam, A. (2023, June 16). *The professions most likely to be paired up in marriage, and more!. The Washington Post*. https://www.washingtonpost.com/business/2023/06/16/what-does-your-job-say-about-whom-youll-marry/ (washingtonpost.com)

Violanti, J. M., & Steege, A. (2021). Law enforcement worker suicide: an updated national assessment. *Policing (Bradford,*

England), *44*(1), 18–31. https://doi.org/10.1108/PIJPSM-09-2019-0157

Violanti, J. M., Vena, J. E., & Marshall, J. R. (1996). Suicides, homicides, and accidental death: a comparative risk assessment of police officers and municipal workers. *American journal of industrial medicine*, *30*(1), 99–104. https://doi.org/10.1002/(SICI)1097-0274(199607)30:1<99::AID-AJIM17>3.0.CO;2-3

ABOUT THE AUTHOR

Shauna 'Doc' Springer, Ph.D., is a licensed psychologist, three-time best-selling author, frequently requested keynote speaker, award-winning podcast host, and one of the world's leading experts on psychological trauma, military transition, suicide prevention, and close relationships. A Harvard graduate who has become a trusted Doc to our nation's military warfighters and first responders, she navigates diverse cultures with exceptional agility.

Doc Springer provides highly specialized expert advising and consultation to public safety leaders, peer-led wellness teams, and organizations that support protectors and defenders through her company, THIN LINE ADVISORY (thinlineadvisory.com). Through this unique relationship of trust and confidence she becomes a force multiplier for high capacity, heart-driven leaders, helping them walk the minefield of challenges associated with leading public safety agencies and teams.

As the Senior Psychological Advisor for Stella Mental Health (stellamentalhealth.com), Dr. Springer has helped advance a groundbreaking approach to addressing psychological trauma that integrates biological and psychological therapies. Stella Mental Health operates clinics in the United States, Australia, and Israel, providing innovative care for post-traumatic stress and other mental health issues.

Doc is also the lead psychological advisor for MAGNUS | ONE (MagnusOne.com) an innovative technology-forward performance eco-system that supports first responders. With the team at MAGNUS | ONE, Doc is helping the wellness conversation away from outdated models towards those aligned with peak performance.

Doc's work is shaped by her unique upbringing, her deep respect for those who serve and protect, and her remarkable insight into human nature. She emphasizes the power of trust and authentic connection as the foundation for transformative growth and the realization of one's full potential.

KEYNOTES, BRIEFINGS AND TRAININGS

Would you like to bring Dr. Shauna Springer in for a **keynote, leadership briefing, or longer training**?

Dr. Springer works with public safety agencies, mental health systems, and organizations that support protectors and defenders. She can deliver one of the topics below or build a customized talk or training course within her scope of expertise to meet the specific needs of your organization. The following topics are examples within a wider range of options – organized by audience, to help you quickly identify what fits your mission.

**Starred trainings would be ideal for front loading into public safety academies.*

MENTAL WARFARE AND SUICIDE RISK

Well-suited for: Public safety, military, veterans, leadership audiences

- **FALL OUT**: Ten Common Beliefs that are Killing Our Bravest and the Truths that Save Them* (summarizes the insights in this book over a 2–4-hour course)

- **Hard Stop**: A New Direction for Preventing Suicide* (Dr. Springer's 2-hour core suicide prevention course)
- **Blind Spot:** Why Suicide Risk in First Responders Hides in Plain Sight*
- **Mental Warfare:** A Tactical Analysis of the Inner Battlefield*
- **Red Herring:** Why PTSD Alone Doesn't Explain Suicide Risk*
- **Trauma is the Smoke, Moral Injury is the Fire**: Conscience, Culture and Conflict in Modern Day Policing*
- **Living in the Red Zone:** How Fight-or-Flight Rewrites Identity, Relationships, and Risk*
- **The Even More Invisible Wounds:** Moral Injury, Shame, and Identity Loss*
- **Postvention After Suicide Loss:** A practical, best-practice guide
- **Fire in the Basement**: Following the Smoke of Invisible Suicide Risk in Firefighters

TRAUMA-INFORMED LEADERSHIP AND PEAK PERFORMANCE

Well-suited for: Public safety leadership, command staff, military leaders

- **Ten Ways Leaders Unintentionally Damage Their Teams During Critical Incidents***
  A trauma-informed look at what goes wrong—and what right looks like

- **ELASTICITY:** Evolving the model for sustaining peak performance in public safety (includes confidential, real-time feedback on nervous system functioning and allostatic load)

- **The Cutting-Edge:** Translating Elite Performance Medicine to Public Safety* (migrating what works in Special Forces to Public Safety – based on articles in the Fraternal Order of Police and Police Chief Magazine)

- **Above the Storm**: Controlled Command Strategies for Unrelenting Pressure

- **Suicide in the Ranks:** Leading Your Team Through Traumatic Grief

- **Carrying the Weight of Grief:** Guiding Teams Through Line-of-Duty Loss

- **Leadership in the Red Zone:** Decision-Making When the Nervous System Is Hijacked

- **The Eight Minute Rule**: Building a Culture of Trust and Connection*
- **SMEAR CAMPAIGNS**: Protecting your Team from Media Misrepresentation
- **Controlling the Story Without Hiding the Truth:** Navigating the Media After Critical Incidents
- **Moving Beyond the Resilience Myth:** Elasticity as a Better Model for Sustainable Performance Under Pressure*
- **Leading through the Noise:** Practical Guidance for Exerting Digital Control
- **CALM is Contagious**: Staying Grounded While the World is in Crisis (for dispatch)*

SUPPORTING PROTECTORS AND DEFENDERS

Best for: Peer teams, clinicians, support organizations

- **THE TRUST FACTOR: Building Authentic Connection with Veterans & First Responders**
- **Beyond Trauma:** Unmasking the Deeper Wounds of Protectors and Defenders (to help peer teams and clinicians understand what fuels distress beyond trauma exposure)
- **Still on Watch:** Grief, identity and healing in military and public safety communities
- **After a Suicide:** Reducing Risk of Collateral Damage When the Unthinkable Happens

- **"Doctors" and "Docs":** Why Credentials Don't Equal Credibility
- **Building Trust Across a Cultural Divide** (for civilians who want to work with Veterans and First Responders)
- **A Loaded Conversation:** Culturally respectful ways to discuss firearms and suicide risk
- **Retirement Without Regret:** A Roadmap for People of Action
- **Not Just Burnout: Addressing the quiet collapse of trauma therapists in a world of continuous crisis** (for clinicians)

MILITARY TRANSITION

Best for: Military units, veteran organizations, transition programs, clinicians

- **Costly Regrets:** Why Smart People Make Bad Decisions After Service
- **Walking the Wasteland:** Navigating Identity Loss and rediscovering your purpose during military transition

CLOSE RELATIONSHIPS

Best for: Anyone, these can be adapted for a range of audiences, including veterans, first responders, and civilians

- **IRREPLACEABLE**: The Evolution of Healthy Relationships across 3 phases

- **Why you SHOULD fight**: How conflict avoidance stunts trust and intimacy
- **How to have a GOOD fight**: 5 key insights for fighting without damaging the one you love (pairs with "Why you SHOULD fight" in a 2- hour training on approaching conflict)
- **Why Good People Have Affairs and How to Heal a Broken Trust** (understanding the risk factors for affairs and how to repair trust when an affair comes to light)
- **Six Ways to Have your Partner's "6"**: How to become irreplaceable to your partner
- **Avoiding Sharks in the Dating Pool:** Tactics for Dodging Regrettable Relationships

Every organization is different. Some need a single, powerful keynote to shift culture. Others need time, depth, and practical tools to change outcomes. Dr. Springer regularly works with organizations to adapt these topics—or build new ones—based on mission, role, and risk exposure.

DOC'S OTHER BOOKS

In BEYOND THE MILITARY: A Leader's Handbook for Warrior Reintegration, with co-author Jason Roncoroni, Doc Springer shines new light on the psychological and relationship challenges associated with military transition. BEYOND THE MILITARY asserts that a successful military transition is not ultimately about finding a new job—it is about reconnecting with your deeper identity, enhancing your closest relationships, and discovering the full potential of life after military service. She and Roncoroni provide a detailed, 400-page roadmap to successful transition through this collective work.

In WARRIOR: How to Support Those Who Protect Us, Doc Springer takes us into the heart of trauma – how it impacts us, and how we can heal. Her insights, gleaned from working with countless warriors who have overcome extreme traumas, are both critical and timely as we emerge from years of societal trauma. WARRIOR has been the focus of collective readings by veteran groups and by thousands of therapists who support service members, veterans and first responders.

In her book, RELENTLESS COURAGE: Winning the Battle Against Frontline Trauma, along with Sergeant (Ret) Michael

Sugrue, Dr. Springer tackles the complexity of trauma with the law enforcement community, uncovers the unspoken barriers, and outlines a path to healing. RELENTLESS COURAGE has been described by Lt. Col. David Grossman, best-selling author of On Killing and On Combat as "one of the most important books of our time."

The book ELASTIC IDENTITY is a key collaboration with Dr. Mitch Javidi and MAGNUS | ONE CEO, Jeff Kingsfield. Integrating psychology, leadership science, and experience, this guide defines Elastic Identity—the ability to flex, adapt, and remain grounded amidst change, pressure, and trauma. In this concise guide, we offer practical advice and effective tools that promote authenticity, encouraging readers to reclaim their sense of self and lead with both integrity and strength.

FALL OUT.

TEN COMMON BELIEFS THAT KILL OUR BRAVEST AND THE TRUTHS THAT CAN SAVE THEM

ISBN: 978-1-7368244-3-6

© 2026 Shauna 'Doc' Springer. All rights reserved. No part of this publication may be reproduced, distributed, or transmitted in any form without prior written permission from the author, except in the case of brief quotations embodied in critical reviews and certain other noncommercial uses permitted by copyright law.

For further information, visit www.thinlineadvisory.com

Made in United States
North Haven, CT
20 June 2026

13158851R10118